Lecture Notes in Computer Science　　10120

Commenced Publication in 1973
Founding and Former Series Editors:
Gerhard Goos, Juris Hartmanis, and Jan van Leeuwen

More information about this series at http://www.springer.com/series/8637

Abdelkader Hameurlain · Josef Küng
Roland Wagner (Eds.)

Transactions on Large-Scale Data- and Knowledge- Centered Systems XXIX

Springer

Editors-in-Chief

Abdelkader Hameurlain
IRIT, Paul Sabatier University
Toulouse
France

Roland Wagner
FAW, University of Linz
Linz
Austria

Josef Küng
FAW, University of Linz
Linz
Austria

ISSN 0302-9743   ISSN 1611-3349   (electronic)
Lecture Notes in Computer Science
ISSN 1869-1994
Transactions on Large-Scale Data- and Knowledge-Centered Systems
ISBN 978-3-662-54036-7   ISBN 978-3-662-54037-4   (eBook)
DOI 10.1007/978-3-662-54037-4

Library of Congress Control Number: 2016958995

Printed on acid-free paper

This Springer imprint is published by Springer Nature
The registered company is Springer-Verlag GmbH Germany
The registered company address is: Heidelberger Platz 3, 14197 Berlin, Germany

Preface

This volume contains four fully revised selected regular papers, covering a wide range of different and very hot topics in the field of data and knowledge management systems. These include optimization and cluster validation processes for entity matching, business intelligence systems, and data profiling in the semantic Web.

We would like to sincerely thank the editorial board and the external reviewers for thoroughly refereeing the submitted papers and ensuring the high quality of this volume.

Special thanks go to Gabriela Wagner for her high availability and her valuable work in the realization of this TLDKS volume.

September 2016

Abdelkader Hameurlain
Josef Küng
Roland Wagner

Organization

Editorial Board

Contents

Contents

Sensitivity - An Important Facet of Cluster Validation Process for Entity Matching Technique

Sumit Mishra, Samrat Mondal[✉], and Sriparna Saha

Department of Computer Science and Engineering,
Indian Institute of Technology Patna, Patna 801103, Bihar, India
{sumitmishra,samrat,sriparna}@iitp.ac.in

Abstract. Cluster validity measure is one of the important components of cluster validation process in which once a clustering arrangement is found, then it is compared with the actual clustering arrangement or gold standard if it is available. For this purpose, different external cluster validity measures (VMs) are available. However, all the measures are not equally good for some specific clustering problem. For example, in entity matching technique, F-measure is a preferably used VM than McNemar index as the former satisfies a given set of desirable properties for entity matching problem. But we have observed that even if all the existing desirable properties are satisfied, then also some of the important differences between two clustering arrangements are not detected by some VMs. Thus we propose to introduce another property, termed as sensitivity, which can be added to the desirable property set and can be used along with the existing set of properties for the cluster validation process. In this paper, the sensitivity property of a VM is formally introduced and then the value of sensitivity is computed using the proposed identity matrix based technique. A comprehensive analysis is made to compare some of the existing VMs and then the suitability of the VMs with respect to the entity matching technique is obtained. Thus, this paper helps to improve the performance of the cluster validation process.

Keywords: Sensitivity · Validity measure · Entity matching · Performance evaluation

1 Introduction

In this era of internet, enormous data are generated day by day. Without identifying the structures in the data, this is of very little or no use. Clustering is one of the technique to identify the structures in the data. Clustering partitions the data into different clusters in such a way that data in the same cluster belong to the same object and data in the different clusters belong to different objects. Various clustering algorithms [3,4,8,11,13–15,17–20,22,24,25,27,30,32–37,39,41,42,44,45] exist in the literature. They all have

© Springer-Verlag GmbH Germany 2016
A. Hameurlain et al. (Eds.): TLDKS XXIX, LNCS 10120, pp. 1–39, 2016.
DOI: 10.1007/978-3-662-54037-4_1

their pros and cons. The question which arises here is how to judge the quality of clusters formed by any clustering algorithm. This process of judging the quality of clusters generated by an algorithm is termed as cluster validation process. This process mainly addresses the two questions - (1) How similar are the solutions of two different algorithms? and (2) In the presence of an available optimal solution: how close is the clustering solution to the optimal one?

In answering these questions, the cluster validity measures (VMs) were introduced. These can judge the similarity between two clustering solutions. In literature various VMs [7,16,21,23,26,31,38,40,43] are proposed for this purpose. These VMs are used for the performance evaluation of the solutions of different problems like entity matching, cancer classification, image processing, protein name detection, sentiment analysis etc. In this paper, we mainly focus on VMs such as - *Precision, Recall, F-measure,* etc. that are used in entity matching problem. New measures are proposed to overcome the limitation of old ones like *Adjusted Rand Index* [43] is developed to overcome the limitations of *Rand Index* [38]. The *Precision* and *Recall* are combined to introduce the *F-measure* [23].

Now among the different VMs that are available it is important to know which one is suitable for a specific problem. In this purpose, Wagner et al. [40] came up with some desirable properties (let's call it DP) of the external validity measure. Some of the important properties are

I. The VM should be independent from the number of clusters as well as number of records.
II. A good VM should not impose any constraint either on cluster size or on number of clusters.
III. There should not be any constraint on the dependencies between the gold standard and obtained partitioning.
IV. The VM should provide non-negative values.
V. The VM should lie between 0 and 1.

A good VM should satisfy this set of properties. However, there are cases when in-spite of satisfying this generic property set or *DP*, a validity measure is not able to differentiate between the two clustering arrangements. The following example illustrates this fact.

Example 1. *Consider* $\mathbb{R} = \{r_1, r_2, \ldots, r_7\}$ *be a set of 7 records from a bibliographic dataset. Let* $\mathcal{C} = \{\{r_1, r_2\}, \{r_3, r_4, r_5, r_6, r_7\}\}$ *be the gold standard. Now we consider two clustering arrangements* $\mathcal{C}_1$ *and* $\mathcal{C}_2$ *which are generated by two different algorithms where* $\mathcal{C}_1 = \{\{r_1\}, \{r_2\}, \{r_3\}, \{r_4, r_5\}, \{r_6, r_7\}\}$ *and* $\mathcal{C}_2 = \{\{r_1, r_2\}, \{r_3\}, \{r_4\}, \{r_5\}, \{r_6, r_7\}\}$. *There will be two clustering structures corresponding to both the arrangement* $\mathcal{C}_1$ *and* $\mathcal{C}_2$. *Now to judge these two clustering solutions we are using F-measure. The value of this measure for both the clustering structures is 4/13.*

In this example, although both $\mathcal{C}_1$ and $\mathcal{C}_2$ are different, but the *F-measure* treats these two as same, i.e. *F-measure* fails to judge the difference between the two arrangements. *F-measure* is not as sensitive as it should be. Thus, for any

VM, sensitivity is also a crucial point to consider. Motivated by this, we propose a new property of VM termed as *sensitivity*.

Now the question arises - how to measure this sensitivity? The naive approach to measure the sensitivity requires to explore all the clustering structures. The total number of clustering structures for a given set of records is very high. Thus, using this approach, it will be a very tedious job to find the sensitivity. So in this paper, we have tried to use a technique which explores a less number of structures and still finds the correct sensitivity. In short, we made the following contributions to the paper.

- We formally introduce the sensitivity property of a VM and also analyze the importance of this property in the cluster validation process.
- For obtaining the sensitivity, we have used the notion of Bell Number and Bell Polynomial to find the possible clustering structures.
- We have proposed a method based on identity matrix which explores a subset of the total clustering structures but still obtains the sensitivity.
- Some heuristics are also proposed to further reduce the explored clustering structures.

Organization of the Paper: Sect. 2 gives the preliminaries required to understand the remaining portion of the paper. The proposed sensitivity and its computation process are presented in Sect. 3. Section 4 provides an elaborate discussion on the proposed identity matrix based approach which explores the required clustering structures for computing the sensitivity. The result and analysis are made in Sect. 5. Finally, Sect. 6 concludes the paper and gives the future direction of the work.

2 Preliminaries

In this section we provide few preliminary notions that are required for understanding our proposed approach.

Let $\mathbb{R} = \{r_1, r_2, \ldots, r_n\}$ be a set of n records for clustering. Let $\mathcal{C} = \{C_1, C_2, \ldots, C_K\}$ and $\mathcal{C}' = \{C_1', C_2', \ldots, C_L'\}$ are the two different partitioning of $\mathbb{R}$ in K and L clusters respectively. The clustering is such that $\cup_{k=1}^K C_k = \cup_{l=1}^L C_l' = \mathbb{R}$ and $C_k \cap C_{k'} = C_l' \cap C_{l'}' = \phi$ for $1 \le k \ne k' \le K$ and $1 \le l \ne l' \le L$. Let n_{kl} be the number of records that are in both cluster C_k and C_l', $1 \le k \le K, 1 \le l \le L$. Also, let n_{k*} denotes the number of records in the k-th cluster C_k, $1 \le k \le K$ while n_{*l} represents the number of records in the l-th cluster C_l', $1 \le l \le L$.

Definition 1 (Contingency Table). *The* contingency table, $M = (n_{kl})$ *of the two partitionings -* $\mathcal{C}$ *and* $\mathcal{C}'$ *is a* $K \times L$ *matrix whose kl-th entry equals the number of elements in the intersection of the clusters* C_k *and* C_l'.

$$n_{kl} = |C_k \cap C_l'|, 1 \le k \le K, 1 \le l \le L$$

The *contingency table* for $\mathcal{C}$ and $\mathcal{C}'$ is shown in Table 1. This contingency table is used to compare two partitioning $\mathcal{C}$ and $\mathcal{C}'$.

Table 1. Notation for the contingency table for comparing two clustering arrangement

c \ c'	C'_1	C'_2	...	C'_L	Sums
C_1	n_{11}	n_{12}	...	n_{1L}	$n_{1\bullet}$
C_2	n_{21}	n_{22}	...	n_{2L}	$n_{2\bullet}$
$\vdots$	$\vdots$	$\vdots$	$\ddots$	$\vdots$	$\vdots$
C_K	n_{K1}	n_{K2}	...	n_{KL}	$n_{K\bullet}$
Sums	$n_{\bullet 1}$	$n_{\bullet 2}$	...	$n_{\bullet L}$	$n_{\bullet\bullet} = n$

Definition 2 (Clustering Arrangement). *Given a set of records, the clustering arrangement gives the different ways in which the records can be clustered into different partitions.*

Example 2. *Let* $\mathbb{R} = \{r_1, r_2, r_3\}$ *is a set of three records. These three records can be clustered in following different ways:* $\{\{r_1, r_2, r_3\}\}$, $\{\{r_1\}, \{r_2, r_3\}\}$, $\{\{r_2\}, \{r_1, r_3\}\}$, $\{\{r_3\}, \{r_1, r_2\}\}$, $\{\{r_1\}, \{r_2\}, \{r_3\}\}$. *So total number of clustering arrangements for three records is 5. In a similar manner, the number of clustering arrangements for four records is 15 and for five records is 52. In clustering arrangement, neither the order of the partitions nor the order of elements within each partition are important. Thus, following are treated as same partition.* $\{\{r_1\}, \{r_2, r_3\}\}$, $\{\{r_2, r_3\}, \{r_1\}\}$, $\{\{r_1\}, \{r_3, r_2\}\}$, $\{\{r_3, r_2\}, \{r_1\}\}$.

Definition 3 (Clustering Structure). *The total ways in which two clustering arrangements can be compared is termed as clustering structure.*

Two clustering arrangements can be compared using contingency table as shown in Table 1.

Example 3. *Let* $\mathbb{R} = \{r_1, r_2, r_3\}$ *is the set of 3 records. Total number of clustering arrangements for these 3 records is 5. Let* $\mathbb{CA}$ *be the set of all the clustering arrangements. Thus* $\mathbb{CA} = \{\{\{r_1, r_2, r_3\}\}, \{\{r_1\}, \{r_2, r_3\}\}, \{\{r_2\}, \{r_1, r_3\}\}, \{\{r_3\}, \{r_1, r_2\}\}, \{\{r_1\}, \{r_2\}, \{r_3\}\}\}$. *Any of the clustering arrangements from* $\mathbb{CA}$ *can be compared with any of the 5 clustering arrangement from* $\mathbb{CA}$ *thus making the clustering structure to 25. The possible clustering structures for these 3 records is shown in Table 2. In this table,* **Arrangement: Gold** *corresponds to the gold standard while* **Arrangement: Obtained** *corresponds to the obtained partitioning.*

Definition 4 (Bell Number). *Bell Number [9] counts the number of different ways to partition a set in non-empty subsets.*

The Bell number satisfies the recurrence relation given in Eq. 1.

$$B_{n+1} = \sum_{k=0}^{n} \binom{n}{k} B_k, \qquad B_0 = B_1 = 1 \qquad (1)$$

n-th Bell Number (B_n) provides the total number of clustering arrangements for n records.

Table 2. Possible clustering structures for 3 records

Serial No.	Clustering Structures	
	Arrangement: Gold	Arrangement: Obtained
1	$\{\{r_1, r_2, r_3\}\}$	$\{\{r_1, r_2, r_3\}\}$
2	$\{\{r_1, r_2, r_3\}\}$	$\{\{r_1\}, \{r_2, r_3\}\}$
3	$\{\{r_1, r_2, r_3\}\}$	$\{\{r_2\}, \{r_1, r_3\}\}$
4	$\{\{r_1, r_2, r_3\}\}$	$\{\{r_3\}, \{r_1, r_2\}\}$
5	$\{\{r_1, r_2, r_3\}\}$	$\{\{r_1\}, \{r_2\}, \{r_3\}\}$
6	$\{\{r_1\}, \{r_2, r_3\}\}$	$\{\{r_1, r_2, r_3\}\}$
7	$\{\{r_1\}, \{r_2, r_3\}\}$	$\{\{r_1\}, \{r_2, r_3\}\}$
8	$\{\{r_1\}, \{r_2, r_3\}\}$	$\{\{r_2\}, \{r_1, r_3\}\}$
9	$\{\{r_1\}, \{r_2, r_3\}\}$	$\{\{r_3\}, \{r_1, r_2\}\}$
10	$\{\{r_1\}, \{r_2, r_3\}\}$	$\{\{r_1\}, \{r_2\}, \{r_3\}\}$
11	$\{\{r_2\}, \{r_1, r_3\}\}$	$\{\{r_1, r_2, r_3\}\}$
12	$\{\{r_2\}, \{r_1, r_3\}\}$	$\{\{r_1\}, \{r_2, r_3\}\}$
13	$\{\{r_2\}, \{r_1, r_3\}\}$	$\{\{r_2\}, \{r_1, r_3\}\}$
14	$\{\{r_2\}, \{r_1, r_3\}\}$	$\{\{r_3\}, \{r_1, r_2\}\}$
15	$\{\{r_2\}, \{r_1, r_3\}\}$	$\{\{r_1\}, \{r_2\}, \{r_3\}\}$
16	$\{\{r_3\}, \{r_1, r_2\}\}$	$\{\{r_1, r_2, r_3\}\}$
17	$\{\{r_3\}, \{r_1, r_2\}\}$	$\{\{r_1\}, \{r_2, r_3\}\}$
18	$\{\{r_3\}, \{r_1, r_2\}\}$	$\{\{r_2\}, \{r_1, r_3\}\}$
19	$\{\{r_3\}, \{r_1, r_2\}\}$	$\{\{r_3\}, \{r_1, r_2\}\}$
20	$\{\{r_3\}, \{r_1, r_2\}\}$	$\{\{r_1\}, \{r_2\}, \{r_3\}\}$
21	$\{\{r_1\}, \{r_2\}, \{r_3\}\}$	$\{\{r_1, r_2, r_3\}\}$
22	$\{\{r_1\}, \{r_2\}, \{r_3\}\}$	$\{\{r_1\}, \{r_2, r_3\}\}$
23	$\{\{r_1\}, \{r_2\}, \{r_3\}\}$	$\{\{r_2\}, \{r_1, r_3\}\}$
24	$\{\{r_1\}, \{r_2\}, \{r_3\}\}$	$\{\{r_3\}, \{r_1, r_2\}\}$
25	$\{\{r_1\}, \{r_2\}, \{r_3\}\}$	$\{\{r_1\}, \{r_2\}, \{r_3\}\}$

Let $\mathbb{CA}$ be the set of B_n clustering arrangements for n records. For comparing two clustering arrangements, one of the clustering arrangement from $\mathbb{CA}$ can be compared with any of the B_n clustering arrangement from $\mathbb{CA}$ so the total number of comparisons will be $B_n \times B_n$. Thus B_n^2 clustering structures are possible. Fig. 1 shows the total number of clustering arrangements and clustering structures for a dataset of size n ($1 \leq n \leq 100$).

Bell Number gives the number of clustering arrangements, but it does not provide any information regarding how the records are arranged in any clustering arrangement. For this, we require Bell Polynomial [10]. The n-th (n is a positive integer) complete Bell Polynomial is given by Eq. 2.

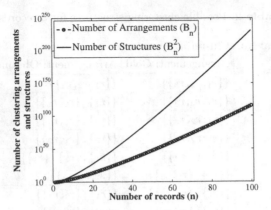

Fig. 1. Possible number of clustering arrangements and structures for n ($1 \leq n \leq 100$) records

$$B_n(x_1, \ldots, x_n) = \sum_{k=1}^{n} B_{n,k}(x_1, x_2, \ldots, x_{n-k+1}) \tag{2}$$

In general
$$B_n(x) = a_1 x + a_2 x^2 + a_3 x^3 + \ldots + a_n x^n \tag{3}$$

The interpretation of Eq. 3 is as follows: there are a_1 numbers of clustering arrangement with single cluster, a_2 clustering arrangements with two clusters, a_3 numbers of clustering arrangement with three clusters and so on. Thus, in general, there are a_n clustering arrangements with n clusters.

Definition 5 (Cluster Validity Measure). *Let* $\mathbb{R} = \{r_1, \ r_2, \ldots, r_n\}$ *be a set of n records and $CA_1, CA_2, \ldots, CA_P$ are P different clustering arrangements obtained by $\mathbb{R}$. Let there is one cluster validity measure M which takes the values $M_1, M_2, \ldots M_P$, then $M_1 \geq M_2 \geq \ldots \geq M_P$ will indicate that $M_1 \uparrow \ldots \uparrow M_P$. Here '$M_i \uparrow M_j$' $\forall i,j \in \{1, 2, \ldots, P\} \wedge i \neq j$ indicates either CA_i is better clustering arrangement than CA_j or both the clustering arrangements are same.*

Some VMs impose a decreasing sequence while some impose an increasing sequence of $CA_1, \ldots, CA_P$. VMs are described with the help of the following notations. *True Positive (TP)* is the number of pairs of records that are in the same cluster in both $\mathcal{C}$ and $\mathcal{C}'$. It is denoted by 'a'. *False Negative (FN)* or 'b' represents the number of pairs of records that are in the same cluster in $\mathcal{C}$ but in different cluster in $\mathcal{C}'$. *False Positive (FP)* or 'c' denotes the number of pairs of records that are in the same cluster in $\mathcal{C}'$, but in different cluster in $\mathcal{C}$. The remaining pair of records means the pair of records that are in different cluster in both $\mathcal{C}$ and $\mathcal{C}'$. This is represented by 'd'. These values are useful to calculate the index values of *counting pairs based measures* [40]. Let the sum of the values of 'a', 'b', 'c' and 'd' is denoted by S. The values of 'a', 'b', 'c', 'd' and S from the contingency Table 1 is computed as follows:

Table 3. Contingency Table for C and C_1

C \ C_1	$\{r_1\}$	$\{r_2\}$	$\{r_3\}$	$\{r_4,r_5\}$	$\{r_6,r_7\}$	Sum
$\{r_1,r_2\}$	1	1	0	0	0	2
$\{r_3,r_4,r_5,r_6,r_7\}$	0	0	1	2	2	5
Sum	1	1	1	2	2	7

Table 4. Contingency Table for C and C_2

C \ C_1	$\{r_1\}$	$\{r_2\}$	$\{r_3\}$	$\{r_4,r_5\}$	$\{r_6,r_7\}$	Sum
$\{r_1,r_2\}$	1	1	0	0	0	2
$\{r_3,r_4,r_5,r_6,r_7\}$	0	0	1	2	2	5
Sum	1	1	1	2	2	7

$$a = \sum_{k,l} \binom{n_{kl}}{2}, \qquad b = \sum_{k} \binom{n_{k*}}{2} - \sum_{k,l} \binom{n_{kl}}{2}, \qquad c = \sum_{l} \binom{n_{*l}}{2} - \sum_{k,l} \binom{n_{kl}}{2}$$
$$d = \binom{n}{2} - (a+b+c), \qquad S = a+b+c+d = \binom{n}{2}$$

3 Sensitivity - A New Property of VM

The formal definition of sensitivity is provided in this section. In this section, we will also discuss why sensitivity is a prime factor and why it is not always the deciding factor.

Definition 6 (Sensitivity of VM). *It refers to the number of unique possible values a validity measure can provide over all the possible clustering structures for a given number of records.*

Let n be the number of records and v be a validity measure. Total number of possible clustering structures for n records is B_n^2. These B_n^2 clustering structures will provide B_n^2 values for a validity measure v. Let out of these B_n^2 values, α values are unique. This α is the sensitivity of a validity measure v for n number of records. Formally, it can be represented as: $s_v(n) = \alpha$.

Let Z is the set of all the possible B_n^2 values for a validity measure v for n number of records. Sensitivity is mapping from Z to α which is shown in Eq. 4.

$$s_v(n) : Z \longrightarrow \alpha \tag{4}$$

Higher value of sensitivity is desirable. The advantage of the high sensitive measure is that it can distinguish two clustering output in a better way.

Example 4. *Total number of clustering structures for 5 records is $= 52 \times 52 = 2704$. As a consequence, the measure also provides 2704 values for all the possible clustering structures. As per Eq. 4, the cardinality of Z is 2704. When the measure is Adjusted Rand Index (ARI), the value of α is 27. The value of α is 11 when the measure is Rand Index (Rand). So, the sensitivity of ARI, i.e. $s_{ARI}(5) = 27$ and sensitivity of Rand Index is $s_{Rand}(5) = 11$. Thus, ARI is more sensitive than Rand Index.*

3.1   Sensitivity: A Prime Factor

Let's consider Example 1 from Sect. 1. The contingency table for first and second clustering structures are given in Table 3 and in Table 4 respectively.

The *F-measure* value for both the contingency table is 4/13. The *Asymmetric F-measure* value for the first contingency table is 88/147 while for the second contingency table it is 34/49. Both the structures are different. But the *F-measure* fails to judge the difference between the two structures, while *Asymmetric F-measure* is able to judge the difference. So *F-measure* is not as sensitive as it should be. Thus, it is important to know how sensitive a validity measure is.

3.2   Sensitivity: Not Always a Deciding Factor

The properties related to the problem domain in which clustering is performed is also very crucial. As an instance, the VM for entity matching should consider the matched and mismatched pairs of records in both the clustering arrangements. There is no need to have a measure for entity matching which does not consider the matching records between the two arrangements. So a VM should also satisfy the conditions related to the domain in which clustering is performed. The following example illustrates this.

Example 5. *Let us consider two VMs - McNemar Index (McI) and F-measure (F) which we want to use for the performance evaluation in entity matching. If the focus is only on the sensitivity, then the McI should be preferred over F-measure because McI is more sensitive than F-measure. However, in practice F-measure is preferred over McI. A measure used for entity matching should consider the matched and mismatched agreement between the two clustering arrangements, but McI does not consider matched agreement. Thus, even if McI has higher sensitivity than F-measure still it does not satisfy the property related to the problem domain. So McI is not preferred over F-measure.*

Similarly for measuring the exactness of binary classification, a measure should consider correctly identified instances and total identified instances. In a similar way, a measure of completeness should consider correctly identified instances and total correct instances. So the properties related to the problem domain in which clustering is performed is also important.

In [28], five indices Rand, Morey and Agresti Adjusted Rand, Hubert and Arabie Adjusted Rand, Jaccard and Fowlkes and Mallows measures were evaluated for measuring agreement between two partitions in hierarchical clustering. In this paper, we have also included the sensitivity along with the existing properties to evaluate the different clustering structures. In most of the cases, a VM with highest sensitivity is selected provided it satisfies the desirable set of properties. Although sensitivity is one of the prime factors, it is not always the deciding factor in the cluster validation process.

3.3   Cluster Validation Process and Sensitivity Computation

To check the effectiveness and performance of a clustering algorithm, we need to form clusters from the given set of records. Once the clusters are formed, different VMs can be applied to see how the clustering algorithm works. Now different VMs will give different results. However, if we know that a particular VM is better than another for a given domain for which clustering is used, then one can use that VM only for performance evaluation purpose. Figure 2 shows the cluster validation process.

Here, we assume that $\mathbb{R} = \{r_1, r_2, \ldots, r_n\}$ be the set of n records in the dataset. A clustering algorithm, say CA is applied to these n records and let's assume that K clusters are formed. For entity matching problem, this K may vary between 1 to n. This clustering result is validated with the help of a validity measure and gold standard (gold standard C_G contains L clusters), i.e., how much similar this result is to the gold standard. In literature, various validity measures exist. Let $M = \{M_1, M_2, \ldots, M_P\}$ be the set of VMs with P elements and only M' ($M' \subseteq M$) satisfies the desirable property set or DP (discussed in Sect. 1) as well the conditions related to the problem domain or Con_{PD} (discussed in Sect. 3.2) in which clustering is performed.

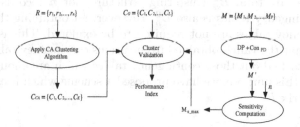

Fig. 2. Cluster Validation process

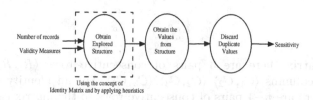

Fig. 3. Sensitivity computation process

Now the sensitivity of all measures in M' is obtained and the measure with highest sensitivity (let $M_{\text{s\_max}}$) is selected. If all other properties of a validity measure is satisfied, then the measure with higher sensitivity is more prominent than that of lower sensitivity.

The proposed method takes as input the number of records and the set of VMs denoted as M'. The sensitivity of these M' VMs is then computed for given number of records. Figure 3 shows the process of sensitivity computation. To obtain the value of sensitivity of a measure, first clustering structures is explored. But if we try to explore all the clustering structures, then it will be very time consuming, so here we propose an approach based on *identity matrix* which explores the very less number of clustering structures as compared to the total number of structures. After exploring the structures, the value of validity measure is obtained for all the clustering structures and unique values are retained. The count of unique values for a validity measure is the sensitivity of that measure. The details of the *identity matrix* based approach are explained in the next section.

4   Proposed Identity Matrix Based Approach

Clustering structure is used to compare two clustering arrangements and these arrangements are compared with the help of contingency tables. Thus, exploring the clustering structures refers to exploring the contingency tables. So here we explore the contingency tables in place of clustering structures. As mentioned in Sect. 2, there are total B_n^2 clustering structures for n records so the total number of contingency tables is also B_n^2. However, for obtaining the sensitivity, all the contingency tables are not required to be explored. This is because, we often observe that the values obtained from various contingency tables are same. Thus we need to find out those contingency tables which are enough to get the sensitivity. For this purpose, we have proposed a scheme which uses the concept of Identity Matrix I.

4.1   Generation of Contingency Tables Using Identity Matrix

For n records an identity matrix of size $n \times n$ is used. As the size of this matrix is $n \times n$ so there will be $n - 1$ pairs of consecutive rows and columns. Such an identity matrix for $n = 4$ is given by Eq. 5.

$$I = \begin{array}{c} \\ R_1 \\ R_2 \\ R_3 \\ R_4 \end{array} \begin{array}{cccc} C_1 & C_2 & C_3 & C_4 \\ \left[\begin{array}{cccc} 1 & 0 & 0 & 0 \\ 0 & 1 & 0 & 0 \\ 0 & 0 & 1 & 0 \\ 0 & 0 & 0 & 1 \end{array} \right] \end{array} \tag{5}$$

In this matrix there are 3 pairs of consecutive rows $\langle R_1, R_2 \rangle, \langle R_2, R_3 \rangle$, $\langle R_3, R_4 \rangle$ and columns $\langle C_1, C_2 \rangle, \langle C_2, C_3 \rangle, \langle C_3, C_4 \rangle$. For an identity matrix I of size $n \times n$, there are $n-1$ pairs of consecutive rows so the matrix can be added in 2^{n-1} different ways row-wise. In a similar way, the matrix can also be added in 2^{n-1} ways column-wise. In the row-wise/column-wise addition, any pair of row/column can be added or not. This can be represented using binary values −1 and 0. '1' represents the addition of the pair of row/column while '0' indicates no addition. Table 5 shows all the binary strings considered for row-wise addition for the identity matrix given in Eq. 5. Table 6 shows all the binary strings considered for column-wise addition for the identity matrix given in Eq. 5. For

Table 5. All possible binary strings of length 3 with their representation and generated matrix row-wise

$\langle R_1, R_2\rangle$	$\langle R_2, R_3\rangle$	$\langle R_3, R_4\rangle$	Binary String $(x \in X)$	Considered rows for Addition	$\mathcal{G}(\mathbf{x}, \mathbf{I})$	Number of rows
0	0	0	000	R_1, R_2, R_3, R_4	$\begin{bmatrix}1\,0\,0\,0\\0\,1\,0\,0\\0\,0\,1\,0\\0\,0\,0\,1\end{bmatrix}$	4
0	0	1	001	$R_1, R_2, R_3 + R_4$	$\begin{bmatrix}1\,0\,0\,0\\0\,1\,0\,0\\0\,0\,1\,1\end{bmatrix}$	3
0	1	0	010	$R_1, R_2 + R_3, R_4$	$\begin{bmatrix}1\,0\,0\,0\\0\,1\,1\,0\\0\,0\,0\,1\end{bmatrix}$	3
0	1	1	011	$R_1, R_2 + R_3 + R_4$	$\begin{bmatrix}1\,0\,0\,0\\0\,1\,1\,1\end{bmatrix}$	2
1	0	0	100	$R_1 + R_2, R_3, R_4$	$\begin{bmatrix}1\,1\,0\,0\\0\,0\,1\,0\\0\,0\,0\,1\end{bmatrix}$	3
1	0	1	101	$R_1 + R_2, R_3 + R_4$	$\begin{bmatrix}1\,1\,0\,0\\0\,0\,1\,1\end{bmatrix}$	2
1	1	0	110	$R_1 + R_2 + R_3, R_4$	$\begin{bmatrix}1\,1\,1\,0\\0\,0\,0\,1\end{bmatrix}$	2
1	1	1	111	$R_1 + R_2 + R_3 + R_4$	$\begin{bmatrix}1\,1\,1\,1\end{bmatrix}$	1

the identity matrix given in Eq. 5, if the binary string used for row-wise addition is '100' then resulting matrix is obtained by adding rows R_1 and R_2. There will be no change in rows R_3 and R_4. Thus the number of rows in resultant matrix will be 3 as opposite to 4 in original matrix I. The resulting matrix is as follows:

$$\begin{array}{c}\\R_1 + R_2\\R_3\\R_4\end{array}\begin{array}{cccc}c_1 & c_2 & c_3 & c_4\\\left[\begin{array}{cccc}1 & 1 & 0 & 0\\0 & 0 & 1 & 0\\0 & 0 & 0 & 1\end{array}\right]\end{array}$$

There are 2^{n-1} combination of 0 and 1 of length $n - 1$. Let X represents the set of all 2^{n-1} binary strings considered for row-wise addition, while Y represents the set of these binary strings considered for column-wise addition. Let $x \in X$ and $y \in Y$ represents any string of 0 and 1 of length $n - 1$. A few functions that we require are defined next.

I. **R** or Reverse operation: $R(x)$ and $R(y)$ represent the reverse of the string x and y respectively. For example, $R(100)$ gives 001.

II. $\mathcal{G}$ or Generate Matrix by Addition operation: $\mathcal{G}(x, I)$ indicates that the rows of I are added according to the string x while $\mathcal{G}(y, I_{row})$ implies the columns of I_{row} are added according to the string y where I_{row} is the matrix obtained by $\mathcal{G}(x, I)$. $\mathcal{G}(x, y, I)$ represents that for generating the contingency table first $\mathcal{G}(x, I)$ is performed which gives I_{row} then $\mathcal{G}(y, I_{row})$ is performed to generate the contingency table M.

Let k be the number of 1's in $x \in X$ and l represent the number of 1's in $y \in Y$. The size of the I_{row} after $\mathcal{G}(x, I)$ will be $(n - k) \times n$. The size of the contingency table M is $n \times (n - l)$ after $\mathcal{G}(y, I)$. The size of the contingency table M after $\mathcal{G}(y, I_{row})$ will be $(n - k) \times (n - l)$. The following example illustrates this operation.

Example 6. *Let $n = 4$. The identity matrix for $n = 4$ is given by Eq. 5. Now the set of 2^{n-1} strings, $X = Y = \{000, 001, 010, 011, 100, 101, 110, 111\}$. The interpretation of these strings when applied on I and considered for row-wise addition is given in Table 5. The interpretation of these strings for column-wise addition when applied on I is shown in Table 6. Consider $x = 010$ and $y = 110$. For generating the contingency table first $\mathcal{G}(x, I)$ is performed which results in I_{row}. Now $\mathcal{G}(y, I_{row})$ is performed to get the contingency table M. The number of 1's in x is 1 while in y is 2. So the size of the contingency table is $(4-1) \times (4-2) = 3 \times 2$. The process of contingency table generation is given next.*

$$
\begin{bmatrix} 1 & 0 & 0 & 0 \\ 0 & 1 & 0 & 0 \\ 0 & 0 & 1 & 0 \\ 0 & 0 & 0 & 1 \end{bmatrix} \xrightarrow{\mathcal{G}(x, I)} \begin{bmatrix} 1 & 0 & 0 & 0 \\ 0 & 1 & 1 & 0 \\ 0 & 0 & 0 & 1 \end{bmatrix} \xrightarrow{\mathcal{G}(y, I_{row})} \begin{bmatrix} 1 & 0 \\ 2 & 0 \\ 0 & 1 \end{bmatrix}
$$
$$
\quad I \qquad\qquad\qquad I_{row} \qquad\qquad\quad M
$$

In short, the process is as follows

$$
\begin{bmatrix} 1 & 0 & 0 & 0 \\ 0 & 1 & 0 & 0 \\ 0 & 0 & 1 & 0 \\ 0 & 0 & 0 & 1 \end{bmatrix} \xrightarrow{\mathcal{G}(x, y, I)} \begin{bmatrix} 1 & 0 \\ 2 & 0 \\ 0 & 1 \end{bmatrix}
$$
$$
\quad I \qquad\qquad\qquad M
$$

The number of 1's in string x and y varies between 0 to $n - 1$ so the size of contingency table varies between 1×1 to $n \times n$. Algorithm 1 is used to generate all the contingency tables using identity matrix based approach. In this algorithm, each combination of row-wise addition is combined with each 2^{n-1} combination of column-wise addition. These two are indicated by line 5 and line 7. Total number of explored contingency tables using the identity matrix based approach is given by Eq. 6. Table 11 in Appendix A shows all the combinations of x and y for 4 records.

$$
\text{Total number of contingency tables} = 2^{n-1} \times 2^{n-1} = 2^{2n-2} \tag{6}
$$

Table 6. All possible binary strings of length 3 with their representation and generated matrix column-wise

$\langle C_1, C_2 \rangle$	$\langle C_2, C_3 \rangle$	$\langle C_3, C_4 \rangle$	Binary String $(y \in Y)$	Considered columns for Addition	$\mathcal{G}(\mathbf{y}, \mathbf{I})$	Number of columns
0	0	0	000	C_1, C_2, C_3, C_4	$\begin{bmatrix} 1 & 0 & 0 & 0 \\ 0 & 1 & 0 & 0 \\ 0 & 0 & 1 & 0 \\ 0 & 0 & 0 & 1 \end{bmatrix}$	4
0	0	1	001	$C_1, C_2, C_3 + C_4$	$\begin{bmatrix} 1 & 0 & 0 \\ 0 & 1 & 0 \\ 0 & 0 & 1 \\ 0 & 0 & 1 \end{bmatrix}$	3
0	1	0	010	$C_1, C_2 + C_3, C_4$	$\begin{bmatrix} 1 & 0 & 0 \\ 0 & 1 & 0 \\ 0 & 1 & 0 \\ 0 & 0 & 1 \end{bmatrix}$	3
0	1	1	011	$C_1, C_2 + C_3 + C_4$	$\begin{bmatrix} 1 & 0 \\ 0 & 1 \\ 0 & 1 \\ 0 & 1 \end{bmatrix}$	2
1	0	0	100	$C_1 + C_2, C_3, C_4$	$\begin{bmatrix} 1 & 0 & 0 \\ 1 & 0 & 0 \\ 0 & 1 & 0 \\ 0 & 0 & 1 \end{bmatrix}$	3
1	0	1	101	$C_1 + C_2, C_3 + C_4$	$\begin{bmatrix} 1 & 0 \\ 1 & 0 \\ 0 & 1 \\ 0 & 1 \end{bmatrix}$	2
1	1	0	110	$C_1 + C_2 + C_3, C_4$	$\begin{bmatrix} 1 & 0 \\ 1 & 0 \\ 1 & 0 \\ 0 & 1 \end{bmatrix}$	2
1	1	1	111	$C_1 + C_2 + C_3 + C_4$	$\begin{bmatrix} 1 \\ 1 \\ 1 \\ 1 \end{bmatrix}$	1

4.2   Complexity of Generating All the Contingency Tables

All the contingency tables are generated from the parameters X, Y and I. As X and Y contain 2^{n-1} binary strings of length $n-1$, the complexity to generate these binary strings is $O((n-1)2^{n-1})$. The time complexity in generating the identity matrix I of size $n \times n$ is $O(n)$ because only diagonal elements are filled with 1. The contingency table is generated when x and y are applied on I, i.e., $\mathcal{G}(x, y, I)$. First $\mathcal{G}(x, I)$ is performed which results in I_{row}. The complexity of obtaining I_{row} from I is $O(n^2)$ because at most $n-1$ pair of rows are added and each row has n elements. After obtaining I_{row}, y is applied on this to generate

---

Algorithm 1. Generate_Contingency_Table

Input: n : Number of records
Output: All the contingency tables
1: $I \leftarrow$ Identity matrix of size $n \times n$
2: $X \leftarrow$ All the possible combination of 0 and 1 of length $n - 1$
3: $Y \leftarrow X$
4: $\mathrm{M} \leftarrow \Phi$ /* M is the set of all contingency tables */
5: **for each** $x \in X$ **do**
6:    $I_{row} \leftarrow \mathcal{G}(x, I)$ /* Perform x on I */
7:    **for each** $y \in Y$ **do**
8:       $M \leftarrow \mathcal{G}(y, I_{row})$ /* Perform y on I_{row} */
9:       $\mathrm{M} \leftarrow \mathrm{M} \cup \{M\}$ /* Add M to the set M */
10: **return** M

---

contingency table M. The complexity involved in obtaining M from I_{row} is also $O(n^2)$ because here at most $n - 1$ pair of columns are added and each column has at most n elements. Thus the complexity in obtaining M from $x \in X$ and $y \in Y$, i.e., $\mathcal{G}(x, y, I)$ is $O(n^2) + O(n^2) \equiv O(n^2)$.

Table 7. Number of 1's vs Number of strings

Number of 1's	0	1	2	$\cdots$	$n - 1$
Number of strings	1	$\binom{n-1}{1}$	$\binom{n-1}{2}$	$\cdots$	$\binom{n-1}{n-1}$

The set of binary strings X and Y are same, so they are generated once. To generate the contingency tables, all possible pairs of $x \in X$ and $y \in Y$ are considered. There are total $2^{n-1} \times 2^{n-1}$ possible pairs of x and y and $O(n^2)$ time is required to generate a contingency table from x, y and I so time complexity to generate all the contingency tables from X, Y and I is given by $O(2^{n-1} \times 2^{n-1} \times n^2)$. Thus the overall complexity becomes $O((n - 1)2^{n-1}) + O(n) + O(2^{n-1} \times 2^{n-1} \times n^2) \equiv O(n^2 2^{2n-2})$.

Let the number of records is n so the length of binary string is $n - 1$. The number of 1's in these binary string will vary between 0 to $n - 1$ as per Table 7.

So in general, the number of binary string containing $i, (0 \le i \le n - 1)$ number of 1's is given by $\binom{n-1}{i}$. If there are i number of 1's in the binary string then either number of row or column will be $n - i$ depending on the binary string considered for row or column. So the number of contingency tables of size $i \times j$ is given by Eq. 7.

$$\text{Number of contingency tables of size } i \times j, 1 \le i, j \le n = \binom{n-1}{n-i} \times \binom{n-1}{n-j}$$
(7)

The size of the contingency table varies from 1×1 to $n \times n$ and total number of contingency tables are 2^{2n-2}. So the relation represented by Eq. 8 can be established.

$$\sum_{i=1}^{n}\sum_{j=1}^{n} \binom{n-1}{n-i} \times \binom{n-1}{n-j} = 2^{2n-2} \tag{8}$$

The overall complexity of the proposed approach for generating all the contingency tables is $O(n^2 2^{2n-2})$ which is exponential in nature. The exponential complexity is overhead in generating all the contingency tables.

The sensitivity of a validity measure for a given number of records does not change. So the sensitivity computation is a one time process. From Algorithm 1, it is evident that the operations in the inner loop are independent of each other, thus the inner loop can be easily parallelized. For speedup the process of sensitivity computation, the parallel version of Algorithm 1 can be used. The Algorithm 2 shows the parallel version of this algorithm. To further reduce the number of explored contingency tables, we have used some of the heuristics. Figure 4 shows the comparison of execution time of serial and parallel version of obtaining sensitivity. The number of CPU cores in our experiment for parallel implementation is 10. From this figure, it is clear that as the number of records increases, the time to compute the sensitivity using parallel version improves significantly compared to the serial version.

Algorithm 2. Generate_Contingency_Table_Parallel

Input: n : Number of records
Output: All the contingency tables
1: $I \leftarrow$ Identity matrix of size $n \times n$
2: $X \leftarrow$ All the possible combination of 0 and 1 of length $n-1$
3: $Y \leftarrow X$
4: $\mathbb{M} \leftarrow \Phi$ /* M is the set of all contingency tables */
5: **for each** $x \in X$ **do**
6:    $I_{row} \leftarrow \mathcal{G}(x, I)$ /* Perform x on I */
   /* **Parallel Section** */
7:    **for each** $y \in Y$ **do**
8:       $M \leftarrow \mathcal{G}(y, I_{row})$ /* Perform y on I_{row} */
9:       $\mathbb{M} \leftarrow \mathbb{M} \cup M$ /* Add M to the set $\mathbb{M}$ */
10: **return** $\mathbb{M}$

4.3 Need for Heuristics

Total number of clustering structures are B_n^2 which is very large. It is clearly shown in Fig. 1 that B_n grows very fast w.r.t. n. To obtain sensitivity, we need to explore B_n^2 structures. However, all the structures may not contribute in computing the sensitivity. Thus, exploring all the B_n^2 structures is very time consuming. To reduce the number of explored structures, identity matrix based approach is

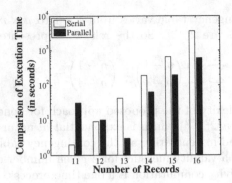

Fig. 4. Comparison of execution time for serial and parallel execution

proposed which explores 2^{2n-2} number of structures as compared to B_n^2. Even 2^{2n-2} is also very high, so to further reduce this we have used some heuristics which do not consider all the strings from set X and Y to generate the contingency tables. Now we describe the notations which are used for understanding the heuristics.

Let the total number of strings in X and Y is represented by n_t, i.e., $n_t = |X| = |Y| = 2^{n-1}$. These X and Y are the combination of palindrome as well as non-palindrome strings. The reason behind visualizing these strings as palindrome and non-palindrome helps us in applying heuristics. One of the heuristic considers the strings as a combination of palindrome and non-palindrome strings. The set of palindrome strings in X and Y is represented by X_p and Y_p respectively. The set of non-palindrome strings in X and Y is represented by X_{np} and Y_{np} respectively. Number of palindrome strings in X and Y are denoted by n_p while n_{np} is used to represent number of non-palindrome strings in X and Y. Thus we can write $X = X_p \cup X_{np}$, $Y = Y_p \cup Y_{np}$ and $n_t = n_p + n_{np}$.

Again X_{np} can be written as $X_{np} = X'_{np} \cup X''_{np}$ where $X'_{np} \cap X''_{np} = \phi \wedge X''_{np} = \{x'' : x'' = R(x')$ where $x' \in X'_{np}\}$. Similarly, $Y_{np} = Y'_{np} \cup Y''_{np}$ where $Y'_{np} \cap Y''_{np} = \phi \wedge Y''_{np} = \{y'' : y'' = R(y')$ where $y' \in Y'_{np}\}$.

Many of the strings in X and Y have their reverse present in the same set. We want to find the set in which no string has their reverse present in the same set. Let X_{wr} and Y_{wr} represent these sets obtained from X and Y respectively. Assume the cardinalities of these sets are n_{wr}. The reverse of the palindrome string is the string itself so all the palindrome strings are included in X_{wr} and Y_{wr}. But half of the strings in X_{np} and Y_{np} have their reverse present in the same set so only half of the strings from X_{np} and Y_{np} are included. Thus $X_{wr} = X_p \cup X'_{np}$ and $Y_{wr} = Y_p \cup Y'_{np}$.

The value of n_{wr} is calculated by Eq. 9. The following example explains the calculation of set X_{wr} and Y_{wr}.

$$n_{wr} = n_p + \frac{1}{2}n_{np} \qquad (9)$$

Example 7. *For number of records* $n = 4$, *the set of binary strings* $X = Y =$ $\{000, 001, 010, 011, 100, 101, 110, 111\}$. $X_p = Y_p = \{000, 010, 101, 111\}$. $X_{np} =$ $Y_{np} = X'_{np} \cup X''_{np} = Y'_{np} \cup Y''_{np} = \{001, 011\} \cup \{100, 110\} = \{001, 011, 100, 110\}$. $n_p = 2^{\frac{4}{2}} = 4$. $n_{np} = 2^{4-1} - 2^{\frac{4}{2}} = 4$. *The half of the strings from* X_{np} *and* Y_{np} *are reverse of another half like the string* 100 *is the reverse of* 001 *and string* 110 *is the reverse of* 011. *So* $X_{wr} = Y_{wr} = \{000, 001, 010, 011, 101, 111\}$. $n_{wr} = n_p + \frac{1}{2}n_{np} = 4 + \frac{1}{2}.4 = 6$.

4.4 First Heuristic (H1)

Statement 1. *The binary strings with all 0's and all 1's are not considered. So* $x, y \notin \{00\ldots0, 11\ldots1\}$.

Reason 1. *All the possible index values generated from all 0's and all 1's combination can also be generated without exploring the contingency tables constructed with all 0's or all 1's.*

Example 8. *For* $n = 4$, *the set of binary strings considered for row and column operation is* $X = Y = \{000, 001, 010, 011, 100, 101, 110, 111\}$. *After considering the heuristic H1 the set of these binary strings will be* $X' = Y' =$ $\{001, 010, 011, 100, 101, 110\}$.

Possible number of explored contingency tables by applying heuristic H1:

- Possible number of rows and columns wise addition $= n_t$
- Possible number of rows and columns wise addition considering heuristic H1 $= n_t - 2$
- Possible number of explored contingency tables by considering heuristic H1 is given by Eq. 10.

$$(n_t - 2) \times (n_t - 2) = n_t^2 - 4n_t + 4 = 2^{2n-2} - 2^{n+1} + 4 \qquad (10)$$

Improvement in the number of contingency tables from 2^{2n-2}:

$$\text{Improvement} = 2^{2n-2} - \left(2^{2n-2} - 2^{n+1} + 4\right) = 2^{n+1} - 4$$

Table 12 in Appendix A shows all the combinations of x and y for 4 records considering heuristic H1.

Values of Validity Measure: Without generating the contingency tables which are constructed by the strings with all 0's and all 1's, the index values can be obtained using the following partition based approach.

Let P be the set of partition of n elements. Let $p_i, 1 \le i \le |P|$ denotes one such partition from the set P. Let p_{max} denotes the maximum element in a partition $p_i, 1 \le i \le |P|$. Let $Pair$ represents the set of the pair which is obtained by the partition elements in P. We discuss this concept of the partition in the following example.

Algorithm 3. Obtain Validity Measure Value - I

Input: P: The set of partition of n elements.

   n: Number of records.

Output: The values of validity measure for matrix size $1 \times i, i \times 1, n \times i, i \times n$.

1: Initialize *Mirkin, Asymm, MH, D* to Φ

2: **for each** $p \in P$ **do**

3:     $mirkin \leftarrow 0, value1 \leftarrow 0, value2 \leftarrow 0$

4:     **for** $(j \leftarrow 0, j < |p|, j++)$ **do**

5:         $mirkin \leftarrow mirkin + p_j^2$

6:         $value1 \leftarrow value1 + p_j(\frac{2p_j}{n+p_j})$

7:         $value2 \leftarrow value2 + p_j(\frac{2}{p_j+1})$

8:     $Mirkin \leftarrow Mirkin \cup (n^2 - mirkin) \cup (mirkin - n)$

9:     $Asymm \leftarrow Asymm \cup (\frac{2p_{max}}{n+p_{max}}) \cup (\frac{1}{n}value1) \cup (\frac{1}{n}value2)$

10:     $MH \leftarrow MH \cup (\frac{1}{n}p_{max}) \cup (\frac{1}{n}|p|)$

11:     $D \leftarrow D \cup (n - p_{max}) \cup (n - |p|)$

Example 9. *Let* $n = 5$. *The partition set* $P = \{\{5\}, \{1, 4\}, \{2, 3\}, \{1, 1, 3\},$ $\{1, 2, \ 2\}, \{1, 1, 1, 2\}, \{1, 1, 1, 1, 1\}\}$ *and pair set* $Pair = \{0, 1, 2, 3, 4, 6, 10\}$. *For a partition* $p_i = \{1, 2, 2\}$ *the* $p_{max} = 2$.

The strings which are not considered in this heuristic will generate the contingency tables of size $1 \times i, i \times 1, n \times i, i \times n, 1 \leq i \leq n$. The contingency tables of these sizes are not explored to generate the values of validity measure. To obtain the values which are generated by the contingency tables of these sizes, partition P and $Pair$ are considered. Algorithm 3 is used to obtain the values of *Mirkin Measure (Mirkin)*, *Asymmetric F-measure (Asymm)*, *Meila-Heckerman Measure (MH)* and *Van Dongen-Measure (D)*. The values of *Precision, Recall, F-measure, Jaccard, Rand Index (Rand), Adjusted Rand Index (ARI), Russel-Rao Index (RRI), Rogers-Tanimoto Index (RTI), Sokal-Sneath Index-1 (SSI1), Sokal-Sneath Index-2 (SSI2), Phi Index (PI), Kulczynski Index (KI), Hubert $\hat{\Gamma}$ Index (HI), Fowlkes-Mallows Index (FM)* and *McNemar index (McI)* is obtained using Algorithm 4. The 'undef' used in Algorithm 4 indicates the mathematically undefined value like 0/0 of the validity measures.

Complexity of generating contingency tables considering heuristic H1: When heuristic H1 is applied $x, y \in \{00 \ldots 0, 11 \ldots 1\}$ is not considered in generating the contingency tables so the cardinality of the set X and Y becomes $2^{n-1} - 2$. Thus the overall complexity will be $O((n - 1)2^{n-1}) + O(n) + O((2^{n-1} - 2) \times (2^{n-1} - 2) \times n^2) \equiv O(n^2(2^{2n-2} - 2^{n+1} + 4)) \equiv O(n^2 2^{2n-2})$.

Complexity of Algorithm 3: This algorithm requires the partition P of n records. Generating P takes exponential time, $O(n \times p(n))$ [12] where $p(n)$ a.k.a. the Hardy and Ramanujam asymptotic formula [5,6], is given by Eq. 11.

$$p(n) = \frac{1}{4n\sqrt{3}} e^{\pi \sqrt{\frac{2n}{3}}} \tag{11}$$

Algorithm 4. Obtain Validity Measure Value - II

Input: *Pair*: The set of pair obtained by partition P.
       S: Total number of pairs from n records.
Output: The values of validity measure for matrix size $1 \times i, i \times 1, n \times i, i \times n$.
1: Initialize *Precision, Recall, Fmeasure, Jaccard, Rand, ARI, HI, FM, KI, SSI1, SSI2,*
    McI, PI, RTI, RRI to Φ
2: **for each** $p \in Pair$ **do**
3:    **if** $p \neq 0$ **then**
4:       $Precision \leftarrow Precision \cup (\frac{p}{S}), Recall \leftarrow Recall \cup (\frac{p}{S})$
5:       $Fmeasure \leftarrow Fmeasure \cup (\frac{2p}{S+p})$
6:       $Jaccard \leftarrow Jaccard \cup (\frac{p}{S}), Rand \leftarrow Rand \cup (\frac{p}{S}) \cup (\frac{S-p}{S})$
7:       $RRI \leftarrow RRI \cup (\frac{p}{S}), RTI \leftarrow RTI \cup (\frac{p}{2S-p}) \cup (\frac{S-p}{S+p})$
8:       $SSI1 \leftarrow SSI1 \cup (\frac{p}{2S-p}), SSI2 \leftarrow SSI2 \cup (\frac{2p}{S+p}) \cup (\frac{2S-2p}{2S-p})$
9:       $KI \leftarrow KI \cup (\frac{S+P}{2S}), FM \leftarrow FM \cup (\sqrt{\frac{P}{S}})$
10:      $McI \leftarrow McI \cup (\sqrt{S-p}) \cup (-\sqrt{S-p}) \cup (\frac{S-2p}{\sqrt{S}})$
11: *Precision* $\leftarrow$ *Precision* $\cup (0) \cup (1) \cup$ (undef), *Recall* $\leftarrow$ *Recall* $\cup (0) \cup (1) \cup$ (undef)
12: *Fmeasure* $\leftarrow$ *Fmeasure* $\cup (1) \cup$ (undef), *Jaccard* $\leftarrow$ *Jaccard* $\cup (0) \cup (1) \cup$ (undef)
13: *Rand* $\leftarrow$ *Rand* $\cup (0) \cup (1)$, *ARI* $\leftarrow$ *ARI* $\cup (0) \cup$ (undef)
14: *RRI* $\leftarrow$ *RRI* $\cup (0)$, *RTI* $\leftarrow$ *RTI* $\cup (0) \cup (1)$
15: *SSI1* $\leftarrow$ *SSI1* $\cup (0) \cup$ (undef), *SSI2* $\leftarrow$ *SSI2* $\cup (0) \cup (1)$
16: *PI* $\leftarrow$ *PI* $\cup$ (undef), *KI* $\leftarrow$ *KI* $\cup$ (undef), *HI* $\leftarrow$ *HI* $\cup$ (undef), *FM* $\leftarrow$ *FM* $\cup$ (undef)
17: *McI* $\leftarrow$ *McI* $\cup$ (undef) $\cup (\sqrt{S}) \cup (-\sqrt{S})$

In Algorithm 3, the outer loop is executed $|P|$ times and its maximum cardinality is $p(n)$ while the inner loop is executed at most n times in each iteration. Thus the time complexity of the algorithm given partition P will be $O(n \times p(n))$. Thus the overall time complexity will be $O(n \times p(n)) + O(n \times p(n)) \equiv O(n \times p(n))$.

Complexity of Algorithm 4: Algorithm 4 requires the *Pair* for n records. The value of pair is calculated for each partition. The time complexity in generating P is $O(n \times p(n))$. For obtaining *Pair* each partition is traversed one time and there are at most n elements in each partition. So the time required to calculate pair for each partition is $O(n)$ and there are at most $p(n)$ such partition. Thus the time complexity of the algorithm given partition P will be $O(n \times p(n))$. Finally the overall time complexity will be $O(n \times p(n)) + O(n \times p(n)) \equiv O(n \times p(n))$.

4.5 Second Heuristic (H2)

Statement 2. *If the contingency table is generated by $\mathcal{G}(x, y, I)$ then the contingency table which is generated by $\mathcal{G}(x', y', I)$ will not be considered where $x' = y$ and $y' = x$. So $\mathcal{G}(x, y, I) \equiv \mathcal{G}(x', y', I)$.*

Reason 2. *The matrix generated by $\mathcal{G}(x, y, I)$ is the transpose of the matrix generated by $\mathcal{G}(x', y', I)$. So the index values provided by both the matrices will be either same or it can be obtained from only one matrix without generating* *the other matrix explicitly so only one is considered. For example; the value of Rand Index will be same for both the matrices while the value of Asymmetric F-measure can not be necessarily same but it can be obtained without generating the transpose matrix explicitly. In general the value of symmetric measure will be same for the matrix and its transpose while the value of asymmetric measure need not be same for the two and the value for transpose matrix can be generated without exploring it explicitly.*

Example 10. *For $n = 4$, the identity matrix I is given by Eq. 5. Let $x = 010$ and $y = 110$. The contingency table generated by $\mathcal{G}(x, y, I)$ is M_{xy} while the contingency table generated by $\mathcal{G}(x', y', I)$ is $M_{x'y'}$. These matrices are transpose of each other.*

$$M_{xy} = \begin{bmatrix} 1 & 0 \\ 2 & 0 \\ 0 & 1 \end{bmatrix} \quad M_{x'y'} = \begin{bmatrix} 1 & 2 & 0 \\ 0 & 0 & 1 \end{bmatrix}$$

For symmetric measures like *F-measure, Jaccard, Rand Index, ARI, Hubert $\hat{\Gamma}$ Index, Fowlkes-Mallows Index, Kulczynski Index, Sokal-Sneath Index, Phi Index, Rogers-Tanimoto Index, Russel-Rao Index, Mirkin Metric, Maximum-Match-Measure* and *Van Dongen-Measure*, the value obtained by one matrix will be same as its transpose. So there is no need to consider the transpose matrix.

For asymmetric measures like *Precision, Recall, McNemar Index, Meila-Heckerman Measure, Asymmetric F-measure*, the value obtained by both the matrices will not be same. In case of *Precision* and *Recall*, the values are interchanged, i.e., the value of *Precision* in one matrix will be the value of *Recall* in its transpose matrix and vice versa. So while obtaining all the values of *Precision*, the *Recall* value is also considered because it will be the *Precision* value in its transpose and we are not considering the transpose matrix. In a similar manner, while obtaining all the values of *Recall*, the *Precision* value is also considered because it will be the *Recall* value in its transpose and we are not considering the transpose matrix.

The values of 'a', 'b', 'c' and 'd' are obtained from the contingency table, i.e., matrix. The values of 'a' and 'd' are same for both matrices and its transpose. However the values of 'b' and 'c' are interchanged, i.e., the value of 'b' in the matrix is the value of 'c' in its transpose and vice versa.

In case of *McNemar Index* the index value obtained by the matrix is different from the value obtained by the transpose matrix so 'c' in the definition of *McNemar Index* is replaced with 'b' to obtain the index values which will be obtained by its transpose and Eq. 12 is used for this purpose.

$$McI = \frac{d - b}{\sqrt{d + b}} \tag{12}$$

In case of *Asymmetric F-measure*, the index value obtained by the matrix is different from the value obtained by the transpose matrix. So to obtain the value of transpose matrix we use Eq. 13. In this way generation of transpose matrix is avoided.

$$Asymmetric\ F\text{-}measure(\mathcal{C}, \mathcal{C}') = \frac{1}{n} \sum_{l=1}^{L} n_{*l} \max_{k=1}^{K} \{F(C_k, C'_l)\} \tag{13}$$

To avoid generation of transpose matrix, Eq. 14 is used for *Meila-Heckerman Measure*.

$$Meila\text{-}Heckerman(\mathcal{C}, \mathcal{C}') = \frac{1}{n} \sum_{l=1}^{L} \max_{C_l \in \mathcal{C}'} \{n_{kl}\} \tag{14}$$

Possible number of explored contingency tables by applying heuristic H2:

- Possible number of rows and columns wise addition $= n_t$
- Possible number of explored contingency tables by considering heuristic H2 is provided by Eq. 15.

$$\frac{n_t \times (n_t + 1)}{2} = \frac{2^{n-1} \times (2^{n-1} + 1)}{2} = 2^{2n-3} + 2^{n-2} \tag{15}$$

Improvement in the number of contingency tables from 2^{2n-2}:

$$Improvement = 2^{2n-2} - \left(2^{2n-3} + 2^{n-2}\right) = 2^{2n-3} - 2^{n-2}$$

Table 13 in Appendix A shows all the combination of x and y for 4 records considering heuristic H2.

Complexity of generating contingency tables considering heuristic H2: When heuristic H2 is applied, all $x \in X$ and all $y \in Y$ are not considered in generating the contingency tables. When H2 is considered total combination will be $(2^{n-1}) + (2^{n-1} - 1) + (2^{n-1} - 2) + \ldots + 1 = \frac{2^{n-1}}{2}(2^{n-1} + 1) = 2^{2n-3} + 2^{n-2}$. Thus the overall complexity becomes $O((n-1)2^{n-1}) + O(n) + O((2^{2n-3} + 2^{n-2}) \times n^2) \equiv O((2^{2n-3} + 2^{n-2}) \times n^2) \equiv O(n^2 2^{2n-3})$.

4.6   Third Heuristic (H3)

Statement 3. *Let $M_{xy} = \mathcal{G}(x, y, I)$ and $M_{rxry} = \mathcal{G}(R(x), R(y), I)$. This heuristic implies that only one of the matrices from M_{xy} and M_{rxry} is considered. So $\mathcal{G}(x, y, I) \equiv \mathcal{G}(R(x), R(y), I)$.*

Reason 3. *M_{xy} and M_{rxry} are same under the row and column transformations. They both provide same values of indices. Let $I_{row} = \mathcal{G}(x, I)$ and $I_{R(row)} = \mathcal{G}((R(x), I)$. The matrix I_{row} and $I_{R(row)}$ are same under row and column transformation. Let $I_{R(row)}^T$ is the matrix after row transformation on $I_{R(row)}$. Let $M_{xy} = \mathcal{G}(y, I_{row})$ and $M_{rxry} = \mathcal{G}(R(y), I_{R(row)})$. M_{xy} and M_{rxry} are same under the row and column transformation. Let $M'_{rxry} = \mathcal{G}(R(y), I_{R(row)}^T)$. The matrix M_{xy} and M'_{rxry} are same under column transformation.*

Example 11. *For $n = 4$, the identity matrix I is given by Eq. 5. Let $x = 011$ and $y = 001$. Let $M_{xy} = \mathcal{G}(x, y, I)$ and $M_{rxry} = \mathcal{G}(R(x), R(y), I)$ is M_{rxry}. M_{xy} can be generated by M_{rxry} with row and column transformation.*

$$I_{row} = \begin{bmatrix} 1 & 0 & 0 & 0 \\ 0 & 1 & 1 & 1 \end{bmatrix} \qquad I_{R(row)} = \begin{bmatrix} 1 & 1 & 1 & 0 \\ 0 & 0 & 0 & 1 \end{bmatrix}$$

$$\underbrace{\begin{bmatrix} 1 & 1 & 1 & 0 \\ 0 & 0 & 0 & 1 \end{bmatrix}}_{I_{R(row)}} \xrightarrow{R_1 \leftrightarrow R_2} \underbrace{\begin{bmatrix} 0 & 0 & 0 & 1 \\ 1 & 1 & 1 & 0 \end{bmatrix}}_{I_{R(row)}^T} \xrightarrow{C_1 \leftrightarrow C_4} \underbrace{\begin{bmatrix} 1 & 0 & 0 & 0 \\ 0 & 1 & 1 & 1 \end{bmatrix}}_{I_{row}}$$

$$M_{xy} = \mathcal{G}(y, I_{row}) = \begin{bmatrix} 1 & 0 & 0 \\ 1 & 1 & 1 \end{bmatrix} \qquad M_{rxry} = \mathcal{G}(R(y), I_{R(row)}) = \begin{bmatrix} 1 & 1 & 1 \\ 0 & 0 & 1 \end{bmatrix}$$

$$M'_{rxry} = \mathcal{G}(R(y), I_{R(row)}^T) = \begin{bmatrix} 0 & 0 & 1 \\ 1 & 1 & 1 \end{bmatrix} \qquad \underbrace{\begin{bmatrix} 1 & 1 & 1 \\ 0 & 0 & 1 \end{bmatrix}}_{M_{rxry}} \xrightarrow{R_1 \leftrightarrow R_2} \begin{bmatrix} 0 & 0 & 1 \\ 1 & 1 & 1 \end{bmatrix} \xrightarrow{C_1 \leftrightarrow C_3} \underbrace{\begin{bmatrix} 1 & 0 & 0 \\ 1 & 1 & 1 \end{bmatrix}}_{M_{xy}}$$

$$\underbrace{\begin{bmatrix} 0 & 0 & 1 \\ 1 & 1 & 1 \end{bmatrix}}_{M'_{rxry}} \xrightarrow{C_1 \leftrightarrow C_3} \underbrace{\begin{bmatrix} 1 & 0 & 0 \\ 1 & 1 & 1 \end{bmatrix}}_{M_{xy}}$$

Possible number of explored contingency tables by applying heuristic H3: This can be divided into two cases. First one with even number of records and other with odd number of records.

Number of Records Is Even. Let number of records $= n$ then $|x| = |y| = n-1(odd)$. The value of $n_t = 2^{n-1}$. The value of $n_p = 2^{\frac{n-1+1}{2}} = 2^{\frac{n}{2}}$ and $n_{np} = 2^{n-1} - 2^{\frac{n}{2}}$. From Eq. 9, the value of n_{wr} is as follows

$$n_{wr} = n_p + \frac{1}{2} n_{np} = 2^{\frac{n}{2}} + \frac{1}{2}\left(2^{n-1} - 2^{\frac{n}{2}}\right) = 2^{n-2} + 2^{\frac{n-2}{2}}$$

Let the possible number of explored contingency tables by considering heuristic H3 is denoted by T_{aE} and is calculated by Eq. 16.

$$T_{aE} = n_p \times n_{wr} + \frac{1}{2} n_{np} \times n_t = 2^{2n-3} + 2^{n-1} \tag{16}$$

Improvement in the number of contingency tables from 2^{2n-2}:

$$\text{Improvement} = 2^{2n-2} - \left(2^{2n-3} + 2^{n-1}\right) = 2^{2n-3} - 2^{n-1}$$

Table 14 in Appendix A shows all the combination of x and y for 4 records considering heuristic H3.

Number of Records is Odd. Let number of records $= n$ then $|x| = |y| = n - 1 (even)$. The value of $n_t = 2^{n-1}$. The value of $n_p = 2^{\frac{n-1}{2}}$ and $n_{np} = 2^{n-1} - 2^{\frac{n-1}{2}}$. From Eq. 9, the value of n_{wr} is as follows

$$n_{wr} = n_p + \frac{1}{2}n_{np} = 2^{\frac{n-1}{2}} + \frac{1}{2}\left(2^{n-1} - 2^{\frac{n-1}{2}}\right) = 2^{n-2} + 2^{\frac{n-3}{2}}$$

Let the possible number of explored contingency tables by considering heuristic H3 is denoted by T_{aO} and is calculated by Eq. 17.

$$T_{aO} = n_p \times n_{wr} + \frac{1}{2}n_{np} \times n_t = 2^{2n-3} + 2^{n-2} \qquad (17)$$

Improvement in the number of contingency tables from 2^{2n-2}:

$$\text{Improvement} = 2^{2n-2} - \left(2^{2n-3} + 2^{n-2}\right) = 2^{2n-3} - 2^{n-2}$$

Table 16 in Appendix A shows all the combination of x and y for 4 records when heuristic H3 is considered.

Complexity of generating contingency tables considering heuristic H3: When heuristic H3 is applied, half of the non-palindrome strings $x \in X$ and $y \in Y$ are not considered. The time complexity of checking whether a string of length $n-1$ is palindrome or not is given by $O(\frac{n-1}{2}) \equiv O(n)$. There are total 2^{n-1} binary string so total time complexity in checking whether all the strings are palindrome or not is given by $O(n2^{n-1})$.

Here we have to check for all the combination of x and y but all x and y are not considered in generating the contingency table. Thus the overall complexity becomes $O((n-1)2^{n-1}) + O(n) + O(n2^{n-1}) + O(2^{n-1} \times 2^{n-1} \times n^2) \equiv O(n^2 2^{2n-2})$.

4.7   Applying H1 and H2 Together

- Possible number of rows and columns wise addition $= n_t$
- Possible number of rows and columns wise addition without considering the string with all 0's and all 1's $= n_t - 2$
- Possible number of explored contingency tables by considering the above heuristics is given by Eq. 18.

$$\frac{(n_t-2) \times (n_t-2+1)}{2} = \frac{(2^{n-1}-2) \times (2^{n-1}-1)}{2} = 2^{2n-3} - 3.2^{n-2} + 1 \quad (18)$$

Improvement in the number of contingency tables from 2^{2n-2}:

$$\text{Improvement} = 2^{2n-2} - \left(2^{2n-3} - 3.2^{n-2} + 1\right) = 2^{2n-3} + 3.2^{n-2} - 1$$

Table 15 in Appendix A shows all the combination of x and y for 4 records when heuristics H1 and H2 are considered.

Complexity of generating contingency tables considering heuristics H1 and H2: When heuristic H1 is applied $x, y \in \{00 \ldots 0, 11 \ldots 1\}$ is not considered in generating the contingency table so the cardinality of the set X and Y becomes $n_t - 2$. When heuristic H1 is combined with H2, the total combination of x and y will be $(2^{n-1} - 2) + (2^{n-1} - 3) + \ldots + 1 = \frac{2^{n-1}-2}{2}(2^{n-1} - 1) = 2^{2n-3} - 2^{n-2} - 2^{n-1} + 1$. Thus the overall complexity becomes $O((n-1)2^{n-1}) + O(n) + O((2^{2n-3} - 2^{n-2} - 2^{n-1} + 1) \times n^2) \equiv O((2^{2n-3} - 2^{n-2} - 2^{n-1} + 1) \times n^2) \equiv O(n^2 2^{2n-3})$.

4.8 Applying H2 and H3 Together

This can be divided into two cases. First one with even number of records and other with odd number of records.

Number of Records Is Even. Let h_{aE} be the number of combination of x and y with x value $00 \ldots 0$ and $11 \ldots 1$ considering heuristics H2 and H3.

$$h_{aE} = 2n_{wr} - 1 = 2\left(2^{n-2} + 2^{\frac{n-2}{2}}\right) - 1 = 2^{n-1} + 2^{\frac{n}{2}} - 1$$

Let h_{bE} be the number of combination of x and y with all x value in $X_p - \{00 \ldots 0, 11 \ldots 1\}$ considering heuristics H2 and H3.

$$h_{bE} = (n_p - 2) \times (n_{wr} - 2) = \left(2^{\frac{n}{2}} - 2\right)\left(2^{n-2} + 2^{\frac{n-2}{2}} - 2\right) = 2^{\frac{3n-4}{2}} - 3.2^{\frac{n}{2}} + 4$$

Let h_{cE} be the number of combination of x and y with all x value in $X_{wr} - X_p$ considering heuristics H2 and H3.

$$h_{cE} = \frac{1}{2}n_{np}n_{np} - \left[\left(\frac{n_{np}}{2} - 1\right) + \left(\frac{n_{np}}{2} - 2\right) + \cdots + 1\right] = P - Q$$

$$P = \frac{1}{2}n_{np}n_{np} = \frac{1}{2}\left(2^{n-1} - 2^{\frac{n}{2}}\right)\left(2^{n-1} - 2^{\frac{n}{2}}\right) = 2^{2n-3} - 2^{\frac{3n-2}{2}} + 2^{n-1}$$

$$Q = \left(\frac{n_{np}}{2} - 1\right) + \left(\frac{n_{np}}{2} - 2\right) + \cdots + 1 = \frac{1}{2}\left(\frac{n_{np}}{2} - 1\right)\left[2 + \left(\frac{n_{np}}{2} - 1\right) - 1\right]$$

$$= 2^{2n-5} - 2^{\frac{3n-6}{2}} + 2^{\frac{n-4}{2}}$$

So $h_{cE} = P - Q = \left(2^{2n-3} - 2^{\frac{3n-2}{2}} + 2^{n-1}\right) - \left(2^{2n-5} - 2^{\frac{3n-6}{2}} + 2^{\frac{n-4}{2}}\right)$

$$= 3.2^{2n-5} - 3.2^{\frac{3n-7}{2}} + 2^{n-1} - 2^{\frac{n-4}{2}}$$

Let the possible number of explored contingency tables when heuristics H2 and H3 are applied is T_{bE} and is calculated by Eq. 19.

$$T_{bE} = h_{aE} + h_{bE} + h_{cE}$$

$$= 2^n - 9.2^{\frac{n-4}{2}} + 3.2^{2n-5} - 2^{\frac{3n-6}{2}} + 3 \tag{19}$$

Improvement in the number of contingency tables from 2^{2n-2}:

$$\text{Improvement} = 2^{2n-2} - \left(2^n - 9.2^{\frac{n-4}{2}} + 3.2^{2n-5} - 2^{\frac{3n-6}{2}} + 3\right)$$

$$= 5.2^{2n-5} - 2^n + 9.2^{\frac{n-4}{2}} + 2^{\frac{3n-6}{2}} - 3$$

Table 17 in Appendix A shows all the combination of x and y for 4 records when heuristics H2 and H3 are considered.

Number of Records Is Odd. Let h_{aO} be the number of combination of x and y with x value $00\ldots0$ and $11\ldots1$ considering heuristics H2 and H3.

$$h_{aO} = 2n_{wr} - 1 = 2\left(2^{n-2} + 2^{\frac{n-3}{2}}\right) - 1 = 2^{n-1} + 2^{\frac{n-1}{2}} - 1$$

Let h_{bO} be the number of combination of x and y with all x value in $X_p - \{00\ldots0, 11\ldots1\}$ considering heuristics H2 and H3.

$$h_{bO} = (n_p - 2) \times (n_{wr} - 2) = \left(2^{\frac{n-1}{2}} - 2\right)\left(2^{n-2} + 2^{\frac{n-3}{2}} - 2\right)$$

$$= 2^{\frac{3n-5}{2}} - 2^{n-2} - 3.2^{\frac{n-1}{2}} + 4$$

Let h_{cO} be the number of combination of x and y with all x value in $X_{wr} - X_p$ considering heuristics H2 and H3.

$$h_{cO} = \frac{1}{2}n_{np}n_{np} - \left[\left(\frac{n_{np}}{2} - 1\right) + \left(\frac{n_{np}}{2} - 2\right) + \cdots + 1\right]$$

$$= P - Q$$

$$P = \frac{1}{2}n_{np}n_{np} = \frac{1}{2}\left(2^{n-1} - 2^{\frac{n-1}{2}}\right)\left(2^{n-1} - 2^{\frac{n-1}{2}}\right) = 2^{2n-3} - 2^{\frac{3n-3}{2}} + 2^{n-2}$$

$$Q = \left(\frac{n_{np}}{2} - 1\right) + \left(\frac{n_{np}}{2} - 2\right) + \cdots + 1 = \frac{1}{2}\left(\frac{n_{np}}{2} - 1\right)\left[2 + \left(\frac{n_{np}}{2} - 1\right) - 1\right]$$

$$= 2^{2n-5} - 2^{\frac{3n-7}{2}} - 2^{n-4} + 2^{\frac{n-5}{2}}$$

So $h_{cO} = P - Q$

$$= \left(2^{2n-3} - 2^{\frac{3n-3}{2}} + 2^{n-2}\right) - \left(2^{2n-5} - 2^{\frac{3n-7}{2}} - 2^{n-4} + 2^{\frac{n-5}{2}}\right)$$

$$= 3.2^{2n-5} - 3.2^{\frac{3n-7}{2}} - 2^{\frac{n-5}{2}} + 5.2^{n-4}$$

Let the possible number of explored contingency tables by applying heuristics H2 and H3 are T_{bO} and is calculated by Eq. 20.

$$T_{bO} = h_{aO} + h_{bO} + h_{cO}$$

$$= 9.2^{n-4} - 9.2^{\frac{n-5}{2}} + 3.2^{2n-5} - 2^{\frac{3n-7}{2}} + 3 \qquad (20)$$

Improvement in the number of contingency tables from 2^{2n-2}:

$$\text{Improvement} = 2^{2n-2} - \left(9.2^{n-4} - 9.2^{\frac{n-5}{2}} + 3.2^{2n-5} - 2^{\frac{3n-7}{2}} + 3\right)$$

$$= 5.2^{2n-5} - 9.2^{n-4} + 9.2^{\frac{n-5}{2}} + 2^{\frac{3n-7}{2}} - 3$$

Table 19 in Appendix A shows all the combination of x and y for 5 records when heuristics H2 and H3 are considered.

Complexity of generating contingency tables considering heuristics H2 and H3: When heuristic H3 is applied, half of the non-palindrome strings x and y are not considered. As discussed earlier the time complexity of checking whether the strings are palindrome or not is given by $O(n2^{n-2})$. All the pairs of x and y are checked to generate the contingency table but not all are considered so the overall complexity becomes $O((n-1)2^{n-1}) + O(n) + O(n2^{n-1}) + O(2^{n-1} \times 2^{n-1} \times n^2) \equiv O(n^2 2^{2n-2})$.

4.9   Applying H1, H2 and H3 Together

This can be divided into two cases. First one with even number of records and other with odd number of records.

Number of Records Is Even. Let h_{aE} be the number of combination of x and y with x value $00\ldots0$ and $11\ldots1$ considering heuristics H2 and H3.

$$h_{aE} = 2n_{wr} - 1 = 2^{n-1} + 2^{\frac{n}{2}} - 1$$

From Eq. 19, possible number of matrices explored by applying heuristics H2 and H3 is

$$T_{bE} = 2^n - 9.2^{\frac{n-4}{2}} + 3.2^{2n-5} - 2^{\frac{3n-6}{2}} + 3$$

Let the possible number of explored contingency tables by applying heuristics H1, H2 and H3 is T_E and is calculated as follows

$$T_E = T_{bE} - h_{aE} = 2^{n-1} - 13.2^{\frac{n-4}{2}} + 3.2^{2n-5} - 2^{\frac{3n-6}{2}} + 4$$

Improvement in the number of contingency tables from 2^{2n-2}:

$$\text{Improvement} = 2^{2n-2} - \left(2^{n-1} - 13.2^{\frac{n-4}{2}} + 3.2^{2n-5} - 2^{\frac{3n-6}{2}} + 4\right)$$

$$= 5.2^{2n-5} - 2^{n-1} + 13.2^{\frac{n-4}{2}} + 2^{\frac{3n-6}{2}} - 4$$

Table 18 in Appendix A shows all the combination of x and y for 4 records when heuristics H1, H2 and H3 are considered together.

Number of Records Is Odd. Let h_{aO} be the number of combination of x and y with $x \in \{00\ldots0, 11\ldots1\}$ considering heuristics H2 and H3.

$$h_{aO} = 2n_{wr} - 1 = 2^{n-1} + 2^{\frac{n-1}{2}} - 1$$

From Eq. 20, possible number of matrices explored by applying heuristics H2 and H3 is

$$T_{bO} = 9.2^{n-4} - 9.2^{\frac{n-5}{2}} + 3.2^{2n-5} - 2^{\frac{3n-7}{2}} + 3$$

Let the possible number of explored contingency tables by applying heuristics H1, H2 and H3 is T_O and is calculated as follows

$$T_O = T_{bO} - h_{aO} = 2^{n-4} - 13.2^{\frac{n-5}{2}} + 3.2^{2n-5} - 2^{\frac{3n-7}{2}} + 4$$

Improvement in the number of contingency tables from 2^{2n-2}:

$$\text{Improvement} = 2^{2n-2} - \left(2^{n-4} - 13.2^{\frac{n-5}{2}} + 3.2^{2n-5} - 2^{\frac{3n-7}{2}} + 4\right)$$

$$= 5.2^{2n-5} - 2^{n-4} + 13.2^{\frac{n-5}{2}} + 2^{\frac{3n-7}{2}} - 4$$

Table 20 in Appendix A shows all the combination of x and y for 5 records when heuristics H1, H2 and H3 are considered together.

Complexity of generating contingency tables considering heuristics H1, H2 and H3: When all the three heuristics a applied the overall complexity becomes $O((n-1)2^{n-1}) + O(n) + O(n2^{n-1}) + O((2^{n-1}-2) \times (2^{n-1}-2) \times n^2) \equiv O((2^{n-1}-2) \times (2^{n-1}-2) \times n^2) = O(n^2(2^{2n-2} - 2^{n+1} + 4)) \equiv O(n^2 2^{2n-2})$.

4.10   Performance Improvement

This section shows the performance improvement over 2^{2n-2} explored structures considering all the three heuristics alone as well as together. Figure 5 shows the performance improvement over 2^{2n-2} considering all the three heuristics and their combination. Let performance improvement in terms of percentage is P.I. when all the three heuristics are considered.

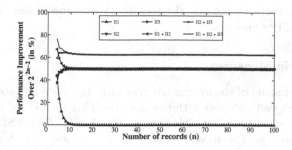

Fig. 5. Performance improvement with all heuristics

Number of Records Is Even:

$$\text{P.I.} = \frac{\text{Improvement in explored structure from } 2^{2n-2}}{2^{2n-2}}$$

$$= \frac{5.2^{2n-5} - 2^{n-1} + 13.2^{\frac{n-4}{2}} + 2^{\frac{3n-6}{2}} - 4}{2^{2n-2}}$$

$$\lim_{n\to\infty} \text{P.I.} = \lim_{n\to\infty} \left(\frac{5}{8} - \frac{1}{2^{n-1}} + \frac{13}{2^{\frac{3n}{2}}} + \frac{1}{2^{\frac{n+2}{2}}} - \frac{1}{2^{2n-4}} \right) = \frac{5}{8}$$

Number of Records Is Odd:

$$\text{P.I.} = \frac{5.2^{2n-5} - 2^{n-4} + 13.2^{\frac{n-5}{2}} + 2^{\frac{3n-7}{2}} - 4}{2^{2n-2}}$$

$$\lim_{n\to\infty} \text{P.I.} = \lim_{n\to\infty} \left(\frac{5}{8} - \frac{1}{2^{n+2}} + \frac{13}{2^{\frac{3n+1}{2}}} + \frac{1}{2^{\frac{n+3}{2}}} - \frac{1}{2^{2n-4}} \right) = \frac{5}{8}$$

When all the three heuristics are considered, the performance improvement in terms of percentage is $\frac{5}{8} \times 100 = 62.5\%$. From Fig. 5, it is clear that when H1 is considered, the improvement is good for small values of n but as n grows the improvement degrades. The improvement reaches to zero with larger value of n. Thus, there is no significant improvement with heuristic H1. The performance improvement is nearly 50% when H2 is considered. With H3 also the improvement is nearly 50%.

The performance improvement is nearly 50% when H1 and H2 are considered together. The improvement is more than 60% when H2 and H3 are considered together. The improvement by considering H2 and H3 alone is 50%, but considering them together further reduces the number of clustering structures. When all the three heuristics are considered together, the performance improvement is about 62.5%. Hence we have to explore only 37.5% clustering structures of 2^{2n-2}. Thus, there is a significant improvement over a number of clustering structures explored for obtaining the sensitivity of validity measure.

5 Result and Analysis

From our proposed approach, we obtain the sensitivity of various measures. The next two subsections provide the sensitivity of some of the symmetric and asymmetric validity measures.

5.1 Symmetric Measures

The sensitivity of some of the symmetric measures is shown in Table 8. Based on the sensitivity we can categorize the symmetric VMs into three groups - *high*, *medium* and *low* sensitive. The VMs such as *Rand, RTI, RRI, SSI2, Mirkin* and *VD* are low sensitive. The measures such as *F-measure (F), Jaccard (J)* and *SSI1* can be considered as medium sensitive and *ARI, PI, KI, HI, FM* may be grouped in high sensitive category.

Table 8. Sensitivity for *F-measure, Jaccard, Rand, ARI, RTI, RRI, PI, SSI1, SSI2, KI, HI, FM, Mirkin* and *VD* for Entity Matching

n	Sensitivity of some of the symmetric validity measures													
	Low						Medium			High				
	VD	RRI	Rand	RTI	SSI2	Mirkin	F	J	SSI1	FM	ARI	KI	PI	HI
3	4	3	4	4	4	4	4	4	4	4	4	4	3	3
4	6	5	7	7	7	7	8	8	8	7	8	8	11	9
5	7	7	11	11	11	11	16	16	16	18	27	19	31	26
6	9	9	16	16	16	16	35	35	35	47	60	47	74	64
7	9	13	22	22	22	22	65	65	65	116	150	120	178	172
8	11	18	29	29	29	29	120	120	120	250	357	277	404	369
9	12	21	37	37	37	37	193	193	193	423	602	474	727	713
10	13	27	46	46	46	46	320	320	320	796	1232	908	1464	1458
11	14	34	56	56	56	56	491	491	491	1777	2530	2080	2974	2915
12	16	39	67	67	67	67	713	713	713	2862	3682	3407	4723	4712
13	16	46	79	79	79	79	1011	1011	1011	4428	5954	5493	8009	8008
14	18	54	92	92	92	92	1453	1453	1453	7149	10957	8882	13011	12706
15	19	61	106	106	106	106	1949	1949	1949	11574	15205	14305	19124	19410
16	20	72	121	121	121	121	2561	2561	2561	18111	21669	22649	29564	30800

The commonly used measure for entity matching is *F-measure* which is the harmonic mean of *Precision* and *Recall*. However, this is a medium sensitive measure. The *FM* measure is the geometric mean of *Precision* and *Recall* and it is a high sensitive measure. So *FM* seems to be a good option for entity matching.

Now let's analyze the other high sensitive measures. Among the high sensitive measures, *KI* is more sensitive than *ARI*. *KI* is the arithmetic mean of *Precision* and *Recall*. In spite of having high sensitivity, *KI* is not suitable as it does not capture the fact that a (50%, 50%) system in terms of *Precision* and *Recall* is often considered better than an (80%, 20%) system. *ARI, PI* and *HI* provide negative index values. The range of these three measures are not clear which makes the interpretation of the index values provided by these measures difficult. Like an index value $-1/81$ provided by *PI* is hard to interpret until its range is not known. So although commonly used validity measure for entity matching is *F-measure*, but our analysis on some existing measure reveals that *FM* will be more suitable for performance evaluation.

Let's consider an example to establish the suitability of *FM* over *F-measure*. Consider $\mathbb{R} = \{r_1, r_2, \ldots, r_8\}$ be the set of 8 records. Let $\mathcal{C} = \{\{r_1, r_2\}, \{r_3\}, \{r_4, r_5, r_6\}, \{r_7\}, \{r_8\}\}$ be the gold standard. Let there are two clustering arrangements $\mathcal{C}_1$ and $\mathcal{C}_2$ such that $\mathcal{C}_1 = \{\{r_1, r_2, r_3, r_4, r_5\}, \{r_6\}, \{r_7\}, \{r_8\}\}$ and $\mathcal{C}_2 = \{\{r_1, r_2, r_3\}, \{r_4\}, \{r_5\}, \{r_6\}, \{r_7\}, \{r_8\}\}$. There are two clustering structures corresponding to both the clustering arrangement $\mathcal{C}_1$ and $\mathcal{C}_2$. The contingency tables corresponding to both the structures are given next.

$$
\begin{array}{c}
\begin{array}{cccc}
r_1r_2r_3r_4r_5 & r_6 & r_7 & r_8
\end{array} \\
\begin{array}{c}
r_1r_2 \\ r_3 \\ r_4r_5r_6 \\ r_7 \\ r_8
\end{array}
\left[
\begin{array}{cccc}
2 & 0 & 0 & 0 \\
1 & 0 & 0 & 0 \\
2 & 1 & 0 & 0 \\
0 & 0 & 1 & 0 \\
0 & 0 & 0 & 1
\end{array}
\right] \\
Ct_1
\end{array}
\qquad
\begin{array}{c}
\begin{array}{cccccc}
r_1r_2r_3 & r_4 & r_5 & r_6 & r_7 & r_8
\end{array} \\
\begin{array}{c}
r_1r_2 \\ r_3 \\ r_4r_5r_6 \\ r_7 \\ r_8
\end{array}
\left[
\begin{array}{cccccc}
2 & 0 & 0 & 0 & 0 \\
1 & 0 & 0 & 0 & 0 \\
0 & 1 & 1 & 1 & 0 & 0 \\
0 & 0 & 0 & 0 & 1 & 0 \\
0 & 0 & 0 & 0 & 0 & 1
\end{array}
\right] \\
Ct_2
\end{array}
$$

The *F-measure* values of both the structures are same, i.e. $\frac{2}{7} = 0.285714$. The *F-measure* treats the two arrangement C_1 and C_2 in a similar manner. The *FM* value for first structure is $\frac{1}{\sqrt{10}} = 0.316227$ while the value of the second structure is $\frac{1}{2\sqrt{3}} = 0.288675$. According to *FM* measure, both the arrangements are different. Thus *FM* is able to differentiate these two clustering arrangements in a much better way than that of *F-measure* and the reason for this is that *FM* measure is more sensitive than *F-measure*.

Table 9. The performance of three techniques for entity matching in terms of *F-measure* and *FM* measure. For Xin Dong dataset, the performance of ArnetMiner is not reported because all the records are not present in ArnetMiner

Dataset	DBLP				ArnetMiner				EMTBD			
	P	R	F	FM	P	R	F	FM	P	R	F	FM
Ajay Gupta	74.75	76.29	75.51	75.51	67.27	75.51	71.15	71.27	100.0	95.94	97.93	97.95
Bin Yu	20.88	100.0	34.54	45.69	72.76	57.94	64.51	64.93	100.0	100.0	100.0	100.0
David Jensen	85.14	100.0	91.97	92.27	94.59	81.25	87.41	87.67	100.0	95.83	97.87	97.90
Jim Smith	39.14	100.0	56.26	62.56	70.36	56.69	62.79	63.16	100.0	100.0	100.0	100.0
Md. Hasanuzzaman	45.83	100.0	62.86	67.70	45.83	100.0	62.86	67.70	100.0	89.09	94.23	94.39
Samrat Goswami	100.0	100.0	100.0	100.0	100.0	100.0	100.0	100.0	100.0	88.23	93.75	93.93
Xin Dong	92.58	95.14	93.84	93.85					100.0	94.89	97.38	97.41

We have applied *F-measure* and *FM* in three systems - DBLP [1], Arnet-Miner [2] and EMTBD [29] developed for entity matching technique in bibliographic database. Table 9 shows the performance for seven datasets. We refer these datasets by the corresponding author name. These datasets are obtained from DBLP and ArnetMiner. The records in these datasets are manually labelled to check the performance of the systems against the gold standard. The commonly used measure for entity matching is *F-measure*. But here we have also reported the *FM* measure along with *F-measure*. In this table, the values for *Precision (P)* and *Recall (R)* are also shown along with *F-measure (F)* and *FM*.

From this table it is clear that when *F-measure* is considered as a performance parameter, then the performance of DBLP is better than ArnetMiner for *Ajay Gupta* and *David Jensen* datasets. The *FM* measure also supports this fact that the performance of DBLP is better than ArnetMiner for these two datasets. The ArnetMiner performs better than DBLP for *Bin Yu* and *Jim Smith* datasets when *F-measure* is used as a performance parameter. When *FM* measure is considered for performance parameter in-spite of *F-Measure* then also ArnetMiner performs better than DBLP. The performance of DBLP and Arnet-Miner are same for *Md. Hasanuzzaman* and *Samrat Goswami* datasets when the *F-measure* is used as a performance evaluation parameter. The *FM* measure provides same index value for DBLP and ArnetMiner for *Md. Hasanuzzaman* and *Samrat Goswami* datasets, i.e., according to *FM* measure also the perfor-mance of DBLP and ArnetMiner are same for these two datasets. Compared to DBLP and ArnetMiner, EMTBD performs poorly for the *Samrat Goswami* dataset. The value of *F-measure* and *FM* measure for this dataset is low for EMTBD as compared to DBLP and ArnetMiner. The reason is that nearly at the same time (1 month difference) two papers are published by *Samrat Goswami* without attributing any common value which EMTBD fails to capture. For the *David Jensen* dataset, EMTBD performs better than DBLP and ArnetMiner and the value of *F-measure* and *FM* both for this dataset is higher than DBLP and ArnetMiner. So *FM* measure is able to evaluate the performance of entity matching.

5.2   Asymmetric Measures

The sensitivity of some of the asymmetric measures is shown in Table 10. Like in case of symmetric measures, if we try to categorize the asymmetric measures we find that *MH* measure is low sensitive, *Precision*, *Recall* and *McI* are medium sensitive and *Asymmetric F-measure* is *very high* sensitive measure. The sensi-tivity of *Asymmetric F-measure* is higher than any other measures that we have studied so far.

Now we need to check whether the other properties are also satisfied by *Asymmetric F-measure*. This measure considers the matching records out of the clusters in obtained clustering arrangement and the gold standard. It also consid-ers the *Precision* and *Recall* for each pair of clusters between obtained clustering arrangement and the gold standard. So, among the asymmetric measures, *Asym-metric F-measure* is a good option for performance evaluation of entity matching as it can distinguish clustering structures in a better way. But for bibliographic database, so far, all records are considered to have equal weight. So the use of symmetric measures appears to be sufficient.

The sensitivity of different symmetric and asymmetric validity measures is shown in Fig. 6. We have categorized the measure in four categories based on their sensitivity - *Low, Medium, High* and *Very high*. We have tried to formally categorize the measures. We have computed the slope of the graph of different validity measures between $n = 15$ and $n = 16$. If the slope of the graph between these two points is greater than 89.99727° then the measure is categorized as *Very high* sensitive. If the slope is in between 89.99727° and 89.97143° then the measure is termed is *High* sensitive. Similarly, if the slope is in between 89.97143° and 88.01338° then the measure is termed is *Medium* sensitive. The measure is termed as *Low* sensitive if the slope is less than 88.01338°.

Table 10. Sensitivity for *Precision, Recall, McI, MH* and *Asymmetric F-measure* for Entity Matching

n	Sensitivity of some of the asymmetric validity measures				
	Low	Medium			Very high
	MH	Precision	Recall	McI	Asymmetric F-measure
3	3	4	4	7	6
4	4	6	6	18	15
5	4	13	13	36	41
6	5	23	23	72	89
7	5	44	44	137	202
8	6	77	77	248	452
9	6	102	102	381	967
10	7	170	170	619	2028
11	7	315	315	941	4187
12	8	408	408	1352	8662
13	8	521	521	1907	18133
14	9	773	773	2614	38994
15	9	1048	1048	3484	90613
16	10	1408	1408	4669	230208

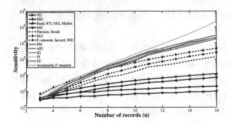

Fig. 6. Sensitivity of different validity measures

Fig. 7. Number of clustering structures explored

5.3   Comparison Among Explored Clustering Structures

Figure 7 shows the total number of clustering structures for entity matching. This number is equal to B_n^2. From this figure, it is clear that there is significant improvement in the number of explored structures for entity matching using the concept of the identity matrix. Explored clustering structures for entity matching when all the three heuristics are applied together is also shown in this figure.

The important thing about obtaining the sensitivity is that it should always be obtained for a large number of records. The reason behind this is that for small values of n, the sensitivity of two measures can be same, but for large value of n the sensitivity may be different. So it gives a wrong impression for sensitivity. From Tables 8 and 10, if we see the sensitivity of *Precision* and *F-measure* for $n = 3$ they are same. But for higher values of n the sensitivity of *F-measure* is higher than that of *Precision*.

6 Conclusion

In this paper we have proposed that sensitivity is another important factor in cluster validation process. Although it is an important property, but it is not a deciding factor all the time. So along with the other properties we can use sensitivity to see the suitability of a cluster validity measure. For computation of sensitivity a huge number of clustering structures are required to explore. But using the proposed identity matrix based approach we explore 2^{2n-2} clustering structures as compared to possible B_n^2 structures. After applying the heuristics, only 37.5% of 2^{2n-2} clustering structures are explored. Thus, it significantly helps in improving the cluster validation process. However, we have observed that by applying the proposed heuristics also some extra contingency tables are explored. So in the future, we would like to study whether any other heuristics or some other technique can further help in reducing the number of explored clustering structures.

A Appendix

Possible combination of strings in X and Y. The strings which are not considered are marked with $*$.

Table 11. Possible strings for $n = 4$

x = 000		x = 001		x = 010		x = 011		x = 100		x = 101		x = 110		x = 111	
x	y	x	y	x	y	x	y	x	y	x	y	x	y	x	y
000	000	001	000	010	000	011	000	100	000	101	000	110	000	111	000
000	001	001	001	010	001	011	001	100	001	101	001	110	001	111	001
000	010	001	010	010	010	011	010	100	010	101	010	110	010	111	010
000	011	001	011	010	011	011	011	100	011	101	011	110	011	111	011
000	100	001	100	010	100	011	100	100	100	101	100	110	100	111	100
000	101	001	101	010	101	011	101	100	101	101	101	110	101	111	101
000	110	001	110	010	110	011	110	100	110	101	110	110	110	111	110
000	111	001	111	010	111	011	111	100	111	101	111	110	111	111	111

Table 12. Possible strings for $n = 4$ by applying H1

x = 000		x = 001		x = 010		x = 011		x = 100		x = 101		x = 110		x = 111	
x	y	x	y	x	y	x	y	x	y	x	y	x	y	x	y
000*	000*	001*	000*	010*	000*	011*	000*	100*	000*	101*	000*	110*	000*	111*	000*
000*	001*	001	001	010	001	011	001	100	001	101	001	110	001	111*	001*
000*	010*	001	010	010	010	011	010	100	010	101	010	110	010	111*	010*
000*	011*	001	011	010	011	011	011	100	011	101	011	110	011	111*	011*
000*	100*	001	100	010	100	011	100	100	100	101	100	110	100	111*	100*
000*	101*	001	101	010	101	011	101	100	101	101	101	110	101	111*	101*
000*	110*	001	110	010	110	011	110	100	110	101	110	110	110	111*	110*
000*	111*	001*	111*	010*	111*	011*	111*	100*	111*	101*	111*	110*	111*	111*	111*

Table 13. Possible Strings for $n = 4$ by applying H2

x = 000		x = 001		x = 010		x = 011		x = 100		x = 101		x = 110		x = 111	
x	y	x	y	x	y	x	y	x	y	x	y	x	y	x	y
000	000	001*	000*	010*	000*	011*	000*	100*	000*	101*	000*	110*	000*	111*	000*
000	001	001	001	010*	001*	011*	001*	100*	001*	101*	001*	110*	001*	111*	001*
000	010	001	010	010	010	011*	010*	100*	010*	101*	010*	110*	010*	111*	010*
000	011	001	011	010	011	011	011	100*	011*	101*	010*	110*	010*	111*	010*
000	100	001	100	010	100	011	100	100	100	101*	100*	110*	100*	111*	100*
000	101	001	101	010	101	011	101	100	101	101	101	110*	101*	111*	101*
000	110	001	110	010	110	011	110	100	110	101	110	110	110	111*	110*
000	111	001	111	010	111	011	111	100	111	101	111	110	111	111	111

Table 14. Possible Strings for $n = 4$ by applying H3

x = 000		x = 001		x = 010		x = 011		x = 100		x = 101		x = 110		x = 111	
x	y	x	y	x	y	x	y	x	y	x	y	x	y	x	y
000	000	001	000	010	000	011	000	100*	000*	101	000	110*	000*	111	000
000	001	001	001	010	001	011	001	100*	001*	101	001	110*	001*	111	001
000	010	001	010	010	010	011	010	100*	010*	101	010	110*	010*	111	010
000	011	001	011	010	011	011	011	100*	011*	101	011	110*	011*	111	011
000*	100*	001	100	010*	100*	011	100	100*	100*	101*	100*	110*	100*	111*	100*
000	101	001	101	010	101	011	101	100*	101*	101	101	110*	101*	111	101
000*	110*	001	110	010*	110*	011	110	100*	110*	101*	110*	110*	110*	111*	110*
000	111	001	111	010	111	011	111	100*	111*	101	111	110*	111*	111	111

Table 15. Possible strings for $n = 4$ by applying heuristics H1 and H2

x = 000		x = 001		x = 010		x = 011		x = 100		x = 101		x = 110		x = 111	
x	y	x	y	x	y	x	y	x	y	x	y	x	y	x	y
000*	000*	001*	000*	010*	000*	011*	000*	100*	000*	101*	000*	110*	000*	111*	000*
000*	001*	001	001	010*	001*	011*	001*	100*	001*	101*	001*	110*	001*	111*	001*
000*	010*	001	010	010	010	011*	010*	100*	010*	101*	010*	110*	010*	111*	010*
000*	011*	001	011	010	011	011	011	100*	011*	101*	011*	110*	011*	111*	011*
000*	100*	001	100	010	100	011	100	100	100	101*	100*	110*	100*	111*	100*
000*	101*	001	101	010	101	011	101	100	101	101	101	110*	101*	111*	101*
000*	110*	001	110	010	110	011	110	100	110	101	110	110	110	111*	110*
000*	111*	001*	111*	010*	111*	011*	111*	100*	111*	101*	111*	110*	111*	111*	111*

Table 16. Possible Strings for $n = 5$ by applying H3

x = 0000		x = 0001		x = 0010		x = 0011		x = 0100		x = 0101		x = 0110		x = 0111	
x	y	x	y	x	y	x	y	x	y	x	y	x	y	x	y
0000	0000	0001	0000	0010	0000	0011	0000	0100*	0000*	0101	0000	0110	0000	0111	0000
0000	0001	0001	0001	0010	0001	0011	0001	0100*	0001*	0101	0001	0110	0001	0111	0001
0000	0010	0001	0010	0010	0010	0011	0010	0100*	0010*	0101	0010	0110	0010	0111	0010
0000	0011	0001	0011	0010	0011	0011	0011	0100*	0011*	0101	0011	0110	0011	0111	0011
0000*	0100*	0001	0100	0010	0100	0011	0100	0100*	0100*	0101	0100	0110*	0100*	0111	0100
0000	0101	0001	0101	0010	0101	0011	0101	0100*	0101*	0101	0101	0110	0101	0111	0101
0000	0110	0001	0110	0010	0110	0011	0110	0100*	0110*	0101	0110	0110	0110	0111	0110
0000	0111	0001	0111	0010	0111	0011	0111	0100*	0111*	0101	0111	0110	0111	0111	0111
0000*	1000*	0001	1000	0010	1000	0011	1000	0100*	1000*	0101	1000	0110*	1000*	0111	1000
0000	1001	0001	1001	0010	1001	0011	1001	0100*	1001*	0101	1001	0110	1001	0111	1001
0000*	1010*	0001	1010	0010	1010	0011	1010	0100*	1010*	0101	1010	0110*	1010*	0111	1010
0000	1011	0001	1011	0010	1011	0011	1011	0100*	1011*	0101	1011	0110	1011	0111	1011
0000*	1100*	0001	1100	0010	1100	0011	1100	0100*	1100*	0101	1100	0110*	1100*	0111	1100
0000*	1101*	0001	1101	0010	1101	0011	1101	0100*	1101*	0101	1101	0110*	1101*	0111	1101
0000*	1110*	0001	1110	0010	1110	0011	1110	0100*	1110*	0101	1110	0110*	1110*	0111	1110
0000	1111	0001	1111	0010	1111	0011	1111	0100*	1111*	0101	1111	0110	1111	0111	1111

x = 1000		x = 1001		x = 1010		x = 1011		x = 1100		x = 1101		x = 1110		x = 1111	
x	y	x	y	x	y	x	y	x	y	x	y	x	y	x	y
1000*	0000*	1001	0000	1010*	0000*	1011	0000	1100*	0000*	1101*	0000*	1110*	0000*	1111	0000
1000*	0001*	1001	0001	1010*	0001*	1011	0001	1100*	0001*	1101*	0001*	1110*	0001*	1111	0001
1000*	0010*	1001	0010	1010*	0010*	1011	0010	1100*	0010*	1101*	0010*	1110*	0010*	1111	0010
1000*	0011*	1001	0011	1010*	0011*	1011	0011	1100*	0011*	1101*	0011*	1110*	0011*	1111	0011
1000*	0100*	1001*	0100*	1010*	0100*	1011	0100	1100*	0100*	1101*	0100*	1110*	0100*	1111*	0100*
1000*	0101*	1001	0101	1010*	0101*	1011	0101	1100*	0101*	1101*	0101*	1110*	0101*	1111	0101
1000*	0110*	1001	0110	1010*	0110*	1011	0110	1100*	0110*	1101*	0110*	1110*	0110*	1111	0110
1000*	0111*	1001	0111	1010*	0111*	1011	0111	1100*	0111*	1101*	0111*	1110*	0111*	1111	0111
1000*	1000*	1001*	1000*	1010*	1000*	1011	1000	1100*	1000*	1101*	1000*	1110*	1000*	1111*	1000*
1000*	1001*	1001	1001	1010*	1001*	1011	1001	1100*	1001*	1101*	1001*	1110*	1001*	1111	1001
1000*	1010*	1001*	1010*	1010*	1010*	1011	1010	1100*	1010*	1101*	1010*	1110*	1010*	1111*	1010*
1000*	1011*	1001	1011	1010*	1011*	1011	1011	1100*	1011*	1101*	1011*	1110*	1011*	1111	1011
1000*	1100*	1001*	1100*	1010*	1100*	1011	1100	1100*	1100*	1101*	1100*	1110*	1100*	1111*	1100*
1000*	1101*	1001*	1101*	1010*	1101*	1011	1101	1100*	1101*	1101*	1101*	1110*	1101*	1111*	1101*
1000*	1110*	1001*	1110*	1010*	1110*	1011	1110	1100*	1110*	1101*	1110*	1110*	1110*	1111*	1110*
1000*	1111*	1001	1111	1010*	1111*	1011	1111	1100*	1111*	1101*	1111*	1110*	1111*	1111	1111

Table 17. Possible Strings for $n = 4$ by applying heuristics H2 and H3

x = 000		x = 001		x = 010		x = 011		x = 100		x = 101		x = 110		x = 111	
x	y	x	y	x	y	x	y	x	y	x	y	x	y	x	y
000	000	001*	000*	010*	000*	011*	000*	100*	000*	101*	000*	110*	000*	111*	000*
000	001	001	001	010*	001*	011	001	100*	001*	101*	001*	110*	001*	111	001
000	010	001	010	010	010	011	010	100*	010*	101	010	110*	010*	111	010
000	011	001*	011*	010*	011*	011	011	100*	011*	101*	011*	110*	011*	111	011
000*	100*	001	100	010*	100*	011	100	100*	100*	101*	100*	110*	100*	111*	100*
000	101	001	101	010	101	011	101	100*	101*	101	101	110*	101*	111	101
000*	110*	001	110	010*	110*	011	110	100*	110*	101*	110*	110*	110*	111*	110*
000	111	001*	111*	010*	111*	011*	111*	100*	111*	101*	111*	110*	111*	111	111

Table 18. Possible Strings for $n = 4$ by applying heuristics H1, H2 and H3

x = 000		x = 001		x = 010		x = 011		x = 100		x = 101		x = 110		x = 111	
x	y	x	y	x	y	x	y	x	y	x	y	x	y	x	y
000*	000*	001*	000*	010*	000*	011*	000*	100*	000*	101*	000*	110*	000*	111*	000*
000*	001*	001	001	010*	001*	011	001	100*	001*	101*	001*	110*	001*	111*	001*
000*	010*	001	010	010	010	011	010	100*	010*	101	010	110*	010*	111*	010*
000*	011*	001*	011*	010*	011*	011	011	100*	011*	101*	011*	110*	011*	111*	011*
000*	100*	001	100	010*	100*	011	100	100*	100*	101*	100*	110*	100*	111*	100*
000*	101*	001	101	010	101	011	101	100*	101*	101	101	110*	101*	111*	101*
000*	110*	001	110	010*	110*	011	110	100*	110*	101*	110*	110*	110*	111*	110*
000*	111*	001*	111*	010*	111*	011*	111*	100*	111*	101*	111*	110*	111*	111*	111*

Table 19. Possible Strings for $n = 5$ by applying heuristics H2 and H3

x = 0000		x = 0001		x = 0010		x = 0011		x = 0100		x = 0101		x = 0110		x = 0111	
x	y	x	y	x	y	x	y	x	y	x	y	x	y	x	y
0000	0000	0001*	0000*	0010*	0000*	0011*	0000*	0100*	0000*	0101*	0000*	0110*	0000*	0111*	0000*
0000	0001	0001	0001	0010	0001	0011	0001	0100*	0001*	0101	0001	0110*	0001*	0111	0001
0000	0010	0001*	0010*	0010	0010	0011	0010	0100*	0010*	0101	0010	0110*	0010*	0111	0010
0000	0011	0001*	0011*	0010*	0011*	0011	0011	0100*	0011*	0101	0011	0110*	0011*	0111	0011
0000*	0100*	0001	0100	0010	0100	0011	0100	0100*	0100*	0101	0100	0110*	0100*	0111	0100
0000	0101	0001*	0101*	0010*	0101*	0011*	0101*	0100*	0101*	0101	0101	0110*	0101*	0111	0101
0000	0110	0001	0110	0010	0110	0011	0110	0100*	0110*	0101	0110	0110	0110	0111	0110
0000	0111	0001*	0111*	0010*	0111*	0011*	0111*	0100*	0111*	0101*	0111*	0110*	0111*	0111	0111
0000*	1000*	0001	1000	0010	1000	0011	1000	0100*	1000*	0101	1000	0110*	1000*	0111	1000
0000	1001	0001	1001	0010	1001	0011	1001	0100*	1001*	0101	1001	0110	1001	0111	1001
0000*	1010*	0001	1010	0010	1010	0011	1010	0100*	1010*	0101	1010	0110*	1010*	0111	1010
0000	1011	0001*	1011*	0010*	1011*	0011*	1011*	0100*	1011*	0101*	1011*	0110*	1011*	0111*	1011*
0000*	1100*	0001	1100	0010	1100	0011	1100	0100*	1100*	0101	1100	0110*	1100*	0111	1100
0000*	1101*	0001	1101	0010	1101	0011	1101	0100*	1101*	0101	1101	0110*	1101*	0111	1101
0000	1110	0001	1110	0010	1110	0011	1110	0100*	1110*	0101	1110	0110*	1110*	0111	1110
0000	1111	0001*	1111*	0010*	1111*	0011*	1111*	0100*	1111*	0101*	1111*	0110*	1111*	0111*	1111*

x = 1000		x = 1001		x = 1010		x = 1011		x = 1100		x = 1101		x = 1110		x = 1111	
x	y	x	y	x	y	x	y	x	y	x	y	x	y	x	y
1000*	0000*	1001*	0000*	1010*	0000*	1011*	0000*	1100*	0000*	1101*	0000*	1110*	0000*	1111*	0000*
1000*	0001*	1001*	0001*	1010*	0001*	1011	0001	1100*	0001*	1101*	0001*	1110*	0001*	1111	0001
1000*	0010*	1001*	0010*	1010*	0010*	1011	0010	1100*	0010*	1101*	0010*	1110*	0010*	1111	0010
1000*	0011*	1001*	0011*	1010*	0011*	1011	0011	1100*	0011*	1101*	0011*	1110*	0011*	1111	0011
1000*	0100*	1001*	0100*	1010*	0100*	1011	0100	1100*	0100*	1101*	0100*	1110*	0100*	1111*	0100*
1000*	0101*	1001*	0101*	1010*	0101*	1011	0101	1100*	0101*	1101*	0101*	1110*	0101*	1111	0101
1000*	0110*	1001	0110	1010*	0110*	1011	0110	1100*	0110*	1101*	0110*	1110*	0110*	1111	0110
1000*	0111*	1001*	0111*	1010*	0111*	1011	0111	1100*	0111*	1101*	0111*	1110*	0111*	1111	0111
1000*	1000*	1001*	1000*	1010*	1000*	1011	1000	1100*	1000*	1101*	1000*	1110*	1000*	1111*	1000*
1000*	1001*	1001	1001	1010*	1001*	1011	1001	1100*	1001*	1101*	1001*	1110*	1001*	1111	1001
1000*	1010*	1001*	1010*	1010*	1010*	1011	1010	1100*	1010*	1101*	1010*	1110*	1010*	1111*	1010*
1000*	1011*	1001*	1011*	1010*	1011*	1011	1011	1100*	1011*	1101*	1011*	1110*	1011*	1111	1011
1000*	1100*	1001*	1100*	1010*	1100*	1011	1100	1100*	1100*	1101*	1100*	1110*	1100*	1111*	1100*
1000*	1101*	1001*	1101*	1010*	1101*	1011	1101	1100*	1101*	1101*	1101*	1110*	1101*	1111*	1101*
1000*	1110*	1001*	1110*	1010*	1110*	1011	1110	1100*	1110*	1101*	1110*	1110*	1110*	1111*	1110*
1000*	1111*	1001*	1111*	1010*	1111*	1011*	1111*	1100*	1111*	1101*	1111*	1110*	1111*	1111	1111

Table 20. Possible strings for $n = 5$ by applying heuristics H1, H2 and H3

x = 0000		x = 0001		x = 0010		x = 0011		x = 0100		x = 0101		x = 0110		x = 0111	
x	y	x	y	x	y	x	y	x	y	x	y	x	y	x	y
0000*	0000*	0001*	0000*	0010*	0000*	0011*	0000*	0100*	0000*	0101*	0000*	0110*	0000*	0111*	0000*
0000*	0001*	0001	0001	0010	0001	0011	0001	0100*	0001*	0101	0001	0110*	0001*	0111	0001
0000*	0010*	0001*	0010*	0010	0010	0011	0010	0100*	0010*	0101	0010	0110*	0010*	0111	0010
0000*	0011*	0001*	0011*	0010*	0011*	0011	0011	0100*	0011*	0101	0011	0110*	0011*	0111	0011
0000*	0100*	0001	0100	0010	0100	0011	0100	0100*	0100*	0101	0100	0110*	0100*	0111	0100
0000*	0101*	0001*	0101*	0010*	0101*	0011*	0101*	0100*	0101*	0101	0101	0110*	0101*	0111	0101
0000*	0110*	0001	0110	0010	0110	0011	0110	0100*	0110*	0101	0110	0110	0110	0111	0110
0000*	0111*	0001*	0111*	0010*	0111*	0011*	0111*	0100*	0111*	0101*	0111*	0110*	0111*	0111	0111
0000*	1000*	0001	1000	0010	1000	0011	1000	0100*	1000*	0101	1000	0110*	1000*	0111	1000
0000*	1001*	0001	1001	0010	1001	0011	1001	0100*	1001*	0101	1001	0110	1001	0111	1001
0000*	1010*	0001	1010	0010	1010	0011	1010	0100*	1010*	0101	1010	0110*	1010*	0111	1010
0000*	1011*	0001*	1011*	0010*	1011*	0011*	1011*	0100*	1011*	0101*	1011*	0110*	1011*	0111*	1011*
0000*	1100*	0001	1100	0010	1100	0011	1100	0100*	1100*	0101	1100	0110*	1100*	0111	1100
0000*	1101*	0001	1101	0010	1101	0011	1101	0100*	1101*	0101	1101	0110*	1101*	0111	1101
0000*	1110*	0001	1110	0010	1110	0011	1110	0100*	1110*	0101	1110	0110*	1110*	0111	1110
0000*	1111*	0001*	1111*	0010*	1111*	0011*	1111*	0100*	1111*	0101*	1111*	0110*	1111*	0111*	1111*

x = 1000		x = 1001		x = 1010		x = 1011		x = 1100		x = 1101		x = 1110		x = 1111	
x	y	x	y	x	y	x	y	x	y	x	y	x	y	x	y
1000*	0000*	1001*	0000*	1010*	0000*	1011*	0000*	1100*	0000*	1101*	0000*	1110*	0000*	1111*	0000*
1000*	0001*	1001*	0001*	1010*	0001*	1011	0001	1100*	0001*	1101*	0001*	1110*	0001*	1111*	0001*
1000*	0010*	1001*	0010*	1010*	0010*	1011	0010	1100*	0010*	1101*	0010*	1110*	0010*	1111*	0010*
1000*	0011*	1001*	0011*	1010*	0011*	1011	0011	1100*	0011*	1101*	0011*	1110*	0011*	1111*	0011*
1000*	0100*	1001*	0100*	1010*	0100*	1011	0100	1100*	0100*	1101*	0100*	1110*	0100*	1111*	0100*
1000*	0101*	1001*	0101*	1010*	0101*	1011	0101	1100*	0101*	1101*	0101*	1110*	0101*	1111*	0101*
1000*	0110*	1001	0110	1010*	0110*	1011	0110	1100*	0110*	1101*	0110*	1110*	0110*	1111*	0110*
1000*	0111*	1001*	0111*	1010*	0111*	1011	0111	1100*	0111*	1101*	0111*	1110*	0111*	1111*	0111*
1000*	1000*	1001*	1000*	1010*	1000*	1011	1000	1100*	1000*	1101*	1000*	1110*	1000*	1111*	1000*
1000*	1001*	1001	1001	1010*	1001*	1011	1001	1100*	1001*	1101*	1001*	1110*	1001*	1111*	1001*
1000*	1010*	1001*	1010*	1010*	1010*	1011	1010	1100*	1010*	1101*	1010*	1110*	1010*	1111*	1010*
1000*	1011*	1001*	1011*	1010*	1011*	1011	1011	1100*	1011*	1101*	1011*	1110*	1011*	1111*	1011*
1000*	1100*	1001*	1100*	1010*	1100*	1011	1100	1100*	1100*	1101*	1100*	1110*	1100*	1111*	1100*
1000*	1101*	1001*	1101*	1010*	1101*	1011	1101	1100*	1101*	1101*	1101*	1110*	1101*	1111*	1101*
1000*	1110*	1001*	1110*	1010*	1110*	1011	1110	1100*	1110*	1101*	1110*	1110*	1110*	1111*	1110*
1000*	1111*	1001*	1111*	1010*	1111*	1011*	1111*	1100*	1111*	1101*	1111*	1110*	1111*	1111*	1111*

References

1. http://dblp.uni-trier.de/
2. https://aminer.org/
3. Aggarwal, C.C.: A survey of stream clustering algorithms. Data Clustering: Algorithms and Applications, p. 231 (2013)
4. Aggarwal, C.C., Zhai, C.: A survey of text clustering algorithms. In: Aggarwal, C.C., Zhai, C. (eds.) Mining Text Data, pp. 77–128. Springer, New York (2012)
5. Andrews, G.E.: The Theory of Partitions, 2nd edn. Cambridge University Press, Cambridge (1998)
6. Andrews, G.E., Eriksson, K.: Integer Partitions. Cambridge University Press, Cambridge (2004)
7. Baker, F.B., Hubert, L.J.: Measuring the power of hierarchical cluster analysis. J. Am. Stat. Assoc. **70**(349), 31–38 (1975)
8. Baudry, J.P., Raftery, A.E., Celeux, G., Lo, K., Gottardo, R.: Combining mixture components for clustering. J. Comput. Graph. Stat. **19**(2) (2010)

9. Becker, H., Riordan, J.: The arithmetic of bell and stirling numbers. Am. J. Math. **70**, 385–394 (1948)
10. Bell, E.T.: Partition polynomials. Ann. Math. **29**, 38–46 (1927)
11. Blachon, S., Pensa, R.G., Besson, J., Robardet, C., Boulicaut, J.F., Gandrillon, O.: Clustering formal concepts to discover biologically relevant knowledge from gene expression data. Silico Biol. **7**(4), 467–483 (2007)
12. Cha, S.H.: Recursive algorithm for generating partitions of an integer. Pace University, Seidenberg School of Computer Science and Information Systems, Technical report (2011)
13. Chen, X., Cai, D.: Large scale spectral clustering with landmark-based representation. In: AAAI (2011)
14. Chen, Y., Sanghavi, S., Xu, H.: Clustering sparse graphs. In: Advances in Neural Information Processing Systems, pp. 2204–2212 (2012)
15. Chitta, R., Jin, R., Jain, A.K.: Efficient kernel clustering using random fourier features. In: 2012 IEEE 12th International Conference on Data Mining (ICDM), pp. 161–170. IEEE (2012)
16. Fowlkes, E.B., Mallows, C.L.: A method for comparing two hierarchical clusterings. J. Am. Stat. Assoc. **78**(383), 553–569 (1983)
17. Girardi, D., Giretzlehner, M., Küng, J.: Using generic meta-data-models for clustering medical data. In: Böhm, C., Khuri, S., Lhotská, L., Renda, M.E. (eds.) ITBAM 2012. LNCS, vol. 7451, pp. 40–53. Springer, Heidelberg (2012). doi:10.1007/978-3-642-32395-9_4
18. Graves, D., Pedrycz, W.: Kernel-based fuzzy clustering and fuzzy clustering: a comparative experimental study. Fuzzy Sets Syst. **161**(4), 522–543 (2010)
19. Guha, S., Rastogi, R., Shim, K.: ROCK: a robust clustering algorithm for categorical attributes. In: 15th International Conference on Data Engineering, 1999, pp. 512–521. IEEE (1999)
20. Höppner, F.: Fuzzy Cluster Analysis: Methods for Classification, Data Analysis and Image Recognition. Wiley, New York (1999)
21. Jackson, D.A., Somers, K.M., Harvey, H.H.: Similarity coefficients: measures of co-occurrence and association or simply measures of occurrence? American Naturalist, pp. 436–453 (1989)
22. Karypis, G., Han, E.H., Kumar, V.: Chameleon: hierarchical clustering using dynamic modeling. Computer **32**(8), 68–75 (1999)
23. Larsen, B., Aone, C.: Fast and effective text mining using linear-time document clustering. In: Proceedings of the Fifth ACM SIGKDD International Conference on Knowledge Discovery and Data Mining, pp. 16–22. ACM (1999)
24. Liu, L., Huang, L., Lai, M., Ma, C.: Projective art with buffers for the high dimensional space clustering and an application to discover stock associations. Neurocomputing **72**(4), 1283–1295 (2009)
25. McNicholas, P.D., Murphy, T.B.: Model-based clustering of microarray expression data via latent gaussian mixture models. Bioinformatics **26**(21), 2705–2712 (2010)
26. Meilă, M., Heckerman, D.: An experimental comparison of model-based clustering methods. Mach. Learn. **42**(1–2), 9–29 (2001)
27. Meng, K., Dong, Z.Y., Wang, D.H., Wong, K.P.: A self-adaptive rbf neural network classifier for transformer fault analysis. IEEE Trans. Power Syst. **25**(3), 1350–1360 (2010)
28. Milligan, G.W., Cooper, M.C.: A study of the comparability of external criteria for hierarchical cluster analysis. Multivar. Behav. Res. **21**(4), 441–458 (1986)

29. Mishra, S., Mondal, S., Saha, S.: Entity matching technique for bibliographic database. In: Decker, H., Lhotská, L., Link, S., Basl, J., Tjoa, A.M. (eds.) DEXA 2013. LNCS, vol. 8056, pp. 34–41. Springer, Heidelberg (2013). doi:10.1007/978-3-642-40173-2_5

30. Müller, K.R., Mika, S., Rätsch, G., Tsuda, K., Schölkopf, B.: An introduction to kernel-based learning algorithms. IEEE Trans. Neural Networks 12(2), 181–201 (2001)

31. Murray, D.A.: Chironomidae: Ecology, Systematics Cytology and Physiology. Elsevier, Amsterdam (1980)

32. Nie, F., Zeng, Z., Tsang, I.W., Xu, D., Zhang, C.: Spectral embedded clustering: a framework for in-sample and out-of-sample spectral clustering. IEEE Trans. Neural Networks 22(11), 1796–1808 (2011)

33. Novák, P., Neumann, P., Macas, J.: Graph-based clustering and characterization of repetitive sequences in next-generation sequencing data. BMC Bioinformatics 11(1), 378 (2010)

34. Pal, N.R., Bezdek, J.C., Tsao, E.C.: Generalized clustering networks and Kohonen's self-organizing scheme. IEEE Trans. Neural Networks 4(4), 549–557 (1993)

35. Pandey, G., Atluri, G., Steinbach, M., Myers, C.L., Kumar, V.: An association analysis approach to biclustering. In: Proceedings of the 15th ACM SIGKDD International Conference on Knowledge Discovery and Data Mining, pp. 677–686. ACM (2009)

36. Park, Y., Moore, C., Bader, J.S.: Dynamic networks from hierarchical Bayesian graph clustering. PloS one 5(1), e8118 (2010)

37. Pensa, R.G., Boulicaut, J.F.: Constrained co-clustering of gene expression data. In: SDM. pp. 25–36. SIAM (2008)

38. Rand, W.M.: Objective criteria for the evaluation of clustering methods. J. Am. Stat. Assoc. 66(336), 846–850 (1971)

39. Schölkopf, B., Smola, A.J.: Learning with Kernels: Support Vector Machines, Regularization, Optimization, and Beyond. MIT press, Cambridge (2002)

40. Wagner, S., Wagner, D.: Comparing clusterings: an overview. Universität Karlsruhe, Fakultät für Informatik Karlsruhe (2007)

41. Wang, X., Davidson, I.: Active spectral clustering. In: IEEE 10th International Conference on Data Mining (ICDM), 2010, pp. 561–568. IEEE (2010)

42. Xiong, H., Steinbach, M., Tan, P.N., Kumar, V.: HICAP: hierarchical clustering with pattern preservation. In: SDM, pp. 279–290 (2004)

43. Yeung, K.Y., Ruzzo, W.L.: Details of the adjusted rand index and clustering algorithms. Bioinformatics 17(9), 763–774 (2001). Supplement to the paper "An empirical study on principal component analysis for clustering gene expression data"

44. Zhang, T., Ramakrishnan, R., Livny, M.: BIRCH: an efficient data clustering method for very large databases. In: ACM SIGMOD Record. vol. 25, pp. 103–114. ACM (1996)

45. Zhuang, X., Huang, Y., Palaniappan, K., Zhao, Y.: Gaussian mixture density modeling, decomposition, and applications. IEEE Trans. Image Process. 5(9), 1293–1302 (1996)

Pay-as-you-go Configuration
of Entity Resolution

Ruhaila Maskat, Norman W. Paton[✉], and Suzanne M. Embury

School of Computer Science, University of Manchester,
Oxford Road, Manchester M13 9PL, UK
maskatr@cs.man.ac.uk, {npaton,suzanne.m.embury}@manchester.ac.uk

Abstract. Entity resolution, which seeks to identify records that represent the same entity, is an important step in many data integration and data cleaning applications. However, entity resolution is challenging both in terms of scalability (all-against-all comparisons are computationally impractical) and result quality (syntactic evidence on record equivalence is often equivocal). As a result, end-to-end entity resolution proposals involve several stages, including *blocking* to efficiently identify candidate duplicates, *detailed comparison* to refine the conclusions from blocking, and *clustering* to identify the sets of records that may represent the same entity. However, the quality of the result is often crucially dependent on configuration parameters in all of these stages, for which it may be difficult for a human expert to provide suitable values. This paper describes an approach in which a complete entity resolution process is optimized, on the basis of feedback (such as might be obtained from crowds) on candidate duplicates. Given such feedback, an evolutionary search of the space of configuration parameters is carried out, with a view to maximizing the fitness of the resulting clusters. The approach is pay-as-you-go in that more feedback can be expected to give rise to better outcomes. An empirical evaluation shows that the co-optimization of the different stages in entity resolution can yield significant improvements over default parameters, even with small amounts of feedback.

1 Introduction

Entity resolution is the task of identifying different records that represent the same entity, and is an important step in many data integration and data cleaning applications [11,21]. A single entity may come to be represented using different records for many reasons; for example, data may be integrated from independently developed sources that have overlapping collections (e.g., different retailers may have overlapping product lines), or a single organization may capture the same data repeatedly (e.g., a police force may encounter the same individual or address many times, in situations where it may be difficult to be confident of the quality of the data). As a result, diverse applications encounter situations in which it is important to ascertain which records refer to the same entity, to allow effective data analysis, cleaning or integration.

© Springer-Verlag GmbH Germany 2016
A. Hameurlain et al. (Eds.): TLDKS XXIX, LNCS 10120, pp. 40–65, 2016.
DOI: 10.1007/978-3-662-54037-4_2

In practice, given large collections, it is impractical to perform a detailed all-against-all comparison, which is $O(n^2)$ on the number of records. As a result, entity resolution tends to involve three principal phases: (i) *blocking*, whereby pairs of records that are candidate duplicates are identified using inexpensive, approximate comparison schemes; (ii) *detailed comparison* in which a distance function compares properties of the candidate duplicates from blocking; and (iii) *clustering*, whereby the records that have been identified as candidate duplicates through blocking are grouped into clusters on the basis of the results of the distance function. There is lots of work on each of these phases; for example, Christen [4] provides a survey of indexing techniques that support blocking, and Hassanzadeh *et al.* [14] compare clustering algorithms that take as input the results of blocking[1].

In an ideal world, off-the-shelf entity resolution techniques would be applied, with minimal manual intervention, to generate dependable results. However, in practice, the quality of the result of entity-resolution techniques is often crucially dependent on configuration parameters in all of *blocking, detailed comparison* and *clustering*. Setting these parameters directly is not straightforward for human experts; the impact of changes to parameters such as thresholds may be wide-ranging and difficult to predict, and there may be subtle inter-dependencies between parameters.

The importance of configuration of entity resolution has been widely recognised, and there are several results on different aspects of the problem. For example, blocking algorithms may use a subset of the fields in a record as the basis for comparison with other records, and apply a function that, given a record, generates one or several blocking keys, each of which gives rise to an index entry; such functions depend on the data to which the algorithm is to be applied, and manual tuning is often both laborious and difficult [4]. Indeed, several results have been reported that seek to automate (often by learning from training data) suitable parameter values for blocking schemes [3,12,26]. It is a similar story for clustering, where algorithms generally make use of thresholds, which also depend on the data that is being clustered. Although there has been some work on the tuning of comparison functions for pairwise matching (e.g. [13,16]), we know of no previous work that seeks to co-optimize the complete lifecycle, from blocking to clustering. The most closely related proposal is probably Corleone [13], which also seeks to optimize the complete entity resolution process, but which differs in optimizing different aspects of an entity resolution pipeline in sequence rather than together, in emphasizing comparison functions rather than wider configuration parameters, and in focusing on pairwise comparison rather than clustering.

In this paper we investigate a pay-as-you-go approach to configuration of a complete entity resolution process. Given feedback on candidate duplicates, the

---

[1] Some entity resolution proposals carry out blocking and pairwise comparison, but stop short of clustering; this is fine up to a point, but clustering proposals are more comprehensive, in that they make the additional decisions as to which groups of candidate duplicates belong together.

approach explores the space of configuration parameters with a view to maximising the quality of the resulting clusters. Thus the entity resolution process is configured automatically in the light of feedback. The approach is pay-as-you-go, in that users can provide as little or as much feedback as they like, reviewing the results in the light of the feedback provided to date, and supplying additional feedback until they are satisfied. An empirical evaluation investigates the trade-off between the amount of feedback provided and the quality of the result.

In this paper, we use feedback that confirms or rejects candidate duplicates to configure the complete entity resolution process. Assume we have an entity resolution system $E(D, P)$, which given a collection of records (represented as tuples in this paper) in a data set D and a set of configuration parameters P (such as similarity thresholds), returns a set of clusters C, such that each cluster $C_i \in C$ is a subset of D. Given feedback on record pairs from D that indicates whether or not the records in the pair are actually duplicates, the problem is to identify parameters P that maximize the quality of the clusters C. Our approach is to use an evolutionary search [24] for parameter values that maximize cluster quality with respect to user feedback, for an existing, state-of-the-art, entity resolution proposal [5] that combines blocking and clustering. As such, this paper is an application of closed-loop evolutionary optimization [19] to entity resolution.

The contributions of the paper are as follows:

1. A generic approach to configuration of parameters for entity resolution that uses feedback on the correctness (or otherwise) of candidate duplicates. These parameters tune the blocking and clustering algorithms, and configure the distance function that is used to compare pairs of records.
2. A description of the application of that approach to produce a self-optimizing pay-as-you-go entity resolution system that uses an evolutionary search over the space of parameter values in which the fitness of alternative sets of parameters is assessed against feedback received on candidate duplicates from blocking.
3. An evaluation of the resulting platform with real world data sets, which shows substantial improvements in cluster quality, even with small amounts of feedback.
4. As the proposal can be considered to be computationally expensive, we describe and evaluate an approach that seeks to retain the results from (3) while also scaling to large data sets.

The paper is structured as follows: Sect. 2 provides the technical context for the remainder of the paper, introducing the terminology and prior results on which later sections build. Section 3 describes our approach to pay-as-you-go entity resolution, which is then evaluated in Sect. 4. An approach to addressing the computational overheads of the approach is presented in Sect. 5. Related work is discussed in Sect. 6, and conclusions follow in Sect. 7.

2   Technical Context

As discussed in the introduction, entity resolution proposals commonly involve two phases, blocking and clustering.

Blocking, given a data set D of type T, associates each element $d_i \in D$ with a set of other elements $M_i \subset D$ such that each element in M_i is a *candidate duplicate* for d_i. In practice, as surveyed by Christen [4], blocking typically yields an index $I : T \rightarrow \{T\}$, where the index may be based on a subset of the attributes of T or some form of string pattern such as n-grams. Requirements of blocking are that it should be efficient, and that M_i should contain most of (preferably all) the actual duplicates of d_i (and preferably not vast numbers of non-duplicates).

Clustering, given a data set D and an index I from blocking, returns a set of clusters C, such that each cluster $C_i \in C$ is a subset of D.

In this paper we do not exhaustively re-review related work on blocking [4] or clustering [14]; the contribution of this paper is on pay-as-you-go configuration of entity resolution, and not on entity resolution techniques *per se*. As such, we do not develop a new entity-resolution proposal, but rather demonstrate our approach on an existing state-of-the-art proposal [5] that is described in this section. We have chosen this proposal because: (i) the blocking phase, in employing a q-gram based hashing scheme, is using an approach that has been shown to be effective in a recent comparison [4]; (ii) the clustering algorithm meets all the generic requirements from [14], in that it is *unconstrained* (the number of clusters is not known in advance), *unsupervised* (clusters can be produced without training data) and *disjoint* (there is no overlap in the membership of clusters); (iii) the clustering algorithm is incremental, enabling new data to be incorporated as it becomes available; (iv) the full entity resolution process, from blocking, through pairwise comparison to clustering is included within a single proposal; and (v) the approach has been shown to be scalable in empirical studies [5]. Although we present our pay-as-you-go configuration approach in the context of this particular proposal, the overall approach is not specific to this technique, and could be applied to configure other approaches.

2.1   Blocking

Blocking, given a data set D of type T, creates an index $I : T \rightarrow \{T\}$ that, given a tuple $d_i \in D$, can retrieve a set of elements $M_i \subset D$ such that each element in M_i is a *candidate duplicate* for d_i. A *candidate duplicate* in this context is a record for which there is some evidence for its equivalence, but where the quality of the comparison may have been traded off for speed. The central questions for blocking are: *what type of index to use* and *how to construct index keys*. In the algorithm of Costa *et al.* [5], hash indexes are used, that associate each tuple d_i with all other tuples that have identical keys, where the keys use an encoding scheme that captures syntactic similarities between the tuples.

Specifically, each tuple may be associated with several keys, each generated using different hash functions, following Algorithm 1. Given a tuple t and configuration parameters P, HASH returns a collection of keys for t. The configuration

---

Algorithm 1. HASH(Tuple t, Parameters P)

---

1 **for** $i \leftarrow$ *1* **to** *P.numKeys* **do**
2 $\quad r \leftarrow \{h_k | a_k \in t, h_k \leftarrow \min\{H_i^1(g)|$
3 $\qquad g \leftarrow$ Q-GRAM$(a_k, P.q)\}\}$
4 $\quad key_i = $ ""
5 $\quad$ **for** $j \leftarrow$ *1* **to** *P.keyComponents* **do**
6 $\qquad key_i \leftarrow key_i ++ \min\{H_j^2(h) | h \in r\}$

7 **return** $< key_1, ..., key_{P.numKeys} >$

---

parameters provide: the number of hash keys to be generated ($P.numKeys$), the number of components in each hash key ($P.keyComponents$) and the size ($P.q$) of q-grams to use for approximate matching. Costa *et al.* assume the presence of two collections of hash functions, H_i^1 and H_j^2, that carry out first and second level hashing (i.e. second level hashing is applied to the result of first level hashing, as described below). These collections of hash functions can be used to generate several keys for each tuple within blocking. HASH proceeds as follows:

- For each of the keys to be generated (line 1) a representation r is created that contains, for each attribute a_k of t, the minimum value obtained when H_i^1 is applied to the q-grams of a_k (lines 2, 3). Thus a single hash code is generated to represent each attribute in t, in such a way that the probability that two values will be assigned the same hash value increases with the overlap between their q-grams.
- The key generated for each tuple is obtained by concatenating (using ++) together $P.keyComponents$ elements from the representation r (line 5). The specific component that contributes to the jth position in the key is the minimum value obtained when H_j^2 is applied to each of the elements in r (line 6).
- The result of the function is a collection of $P.numKeys$ keys (line 7).

To take an example, assume we have two tuples representing person data, each with attributes for forename, surname, street and city: $t_1 = < David, Cameron,$ $10 DowningStreet, London >$ and $t_2 = < Dave, Cameron, DowningStreet,$ $London >$. Assume that the representation r produced in (line 2) for each of the tuples is as follows $r_{t_1} = < h_1, h_2, h_3, h_4 >$, and $r_{t_2} = < k_1, k_2, k_3, k_4 >$, where each h_i and k_j is a hash code. As the surname and city attributes have identical values, $h_2 = k_2$ and $h_4 = k_4$. As the forename and street attributes are similar, and thus have several q-grams in common, there is a good likelihood but no guarantee that $h_1 = k_1$ and that $h_3 = k_3$. Assume that the number of key components is 2, and that the key constructed for t_1 is $H_1^2(h_1) ++ H_2^2(h_2)$. There is then a reasonable likelihood that the first component of the key for t_2 will be the same as for t_1 and a very strong likelihood that the second component of the key will be the same, although it is possible that either or both will be different and thus that the index keys of the tuples will not match.

In constructing the index there is a trade-off: the more index entries there are for a tuple (i.e. the higher is $P.numKeys$), the more the recall of blocking (the

---

Algorithm 2. POPULATECLUSTERS(Clusters C, Index I, Set<Tuple> NewTuples, Parameters P)

---

1 **for** $t \in NewTuples$ **do**
2      $N \leftarrow$ KNEARESTNEIGHBORS(t, I, P)
3      $c \leftarrow$ MOSTLIKELYCLUSTER(N, C, t, P)
4   **if** $c == null$ **then**
5          $newCluster \leftarrow$ **new** $Cluster(t)$;
6          $C \leftarrow C \cup newCluster$
7   **else**
8          $c \leftarrow c \cup t$

9 Return C

---

fraction of the correct tuples that are returned) will rise while its precision (the fraction of the returned tuples that are correct) will fall. By contrast, increasing the number of key components increases the precision of blocking but reduces recall. An empirical evaluation of these features is provided in the original paper on the technique [5].

2.2 Clustering

Given an optional set of existing clusters C, an index from blocking I, a set of tuples $NewTuples$ and some configuration parameters P, POPULATECLUSTERS updates the clusters C to take into account the presence of the $NewTuples$. As a result, this algorithm can be used to cluster an entire data set in one go, or incrementally to cluster data as it becomes available. The top level of the algorithm is given in Algorithm 2, which proceeds as follows:

- For each new tuple t (line 1), its K nearest neighbors are retrieved as N. The function KNEARESTNEIGHBORS (line 2) returns up to P.K entries from the index I that are the closest to t according to a DISTANCE function (described below), and also above a *similarityThreshold* from P.
- The most likely cluster for t is identified by MOSTLIKELYCLUSTER (line 3), which returns a cluster from the clusters of the tuples in N, following a voting procedure involving the neighbors in N. In essence, each neighbor of t, $n_t \in N$, adds a contribution $\frac{1}{Distance(t,n_t)}$ to the score of its cluster, and the new tuple t is considered to be a candidate to be returned by MOSTLIKELYCLUSTER whenever its score from voting exceeds a *membershipThreshold* from P.
- Where a cluster is identified by MOSTLIKELYCLUSTER, t is added to this cluster (line 8), otherwise a new cluster is created with t as its only member (lines 5 and 6).

The DISTANCE function, given two tuples s and t, computes a weighted sum of the n-gram distance (denoted *dist*) of the attributes, so for two tuples $s = < a_{s,1}, ..., a_{s,n} >$ and $t = < a_{t,1}, ..., a_{t,n} >$, their distance is $w_1 \times dist(a_{s,1}, a_{t,1}) + ... + w_n \times dist(a_{s,n}, a_{t,n})$, where the weights are from P.

Algorithm 3. ENTITYRESOLUTION(Set<Tuple> Data, Parameters P)

```
1  I ← new Index();
2  for d ∈ Data do
3  │   Keys = HASH(d, P)
4  │   for k ∈ Keys do
5  │   │   I.Insert(k, d)

6  C ← new Clusters();
7  POPULATECLUSTERS(C, I, Data, P)
8  return C
```

Algorithm 3 provides the top-level pseudo-code of ENTITYRESOLUTION that, given some *Data* to be clustered and control parameters *P*, shows how HASH and POPULATECLUSTERS can be used together to create a new data clustering.

2.3 Configuration Parameters

The algorithms described above make a range of different decisions during both blocking and clustering, where these decisions can be tuned using configuration parameters; the parameters are summarized in Table 1. *The hypothesis that this paper sets out to test is that these parameters can be optimized in the light of feedback to improve the quality of clustering for records for which no feedback has yet been obtained.* Although several of these parameters are strategy-specific, blocking and clustering strategies tend to make similar types of decision, so we suggest that these provide a representative collection of tuning parameters for a study on parameter tuning.

3   Pay-as-you-go Clustering

3.1   Overview

Automatic entity resolution is challenging because duplicate records typically manifest representational heterogeneities and data level inconsistencies that it is difficult for computer systems to unravel. As discussed in general terms in Sect. 1, and as detailed for a specific entity resolution strategy in Sect. 2.3, entity resolution techniques make decisions that are often guided by configuration parameters, the most effective settings for which may be data specific, and which may be challenging to set manually. This section details our approach to optimizing parameter setting, in the light of feedback, where the incremental collection of feedback allows a pay-as-you-go approach.

We cast the search for effective parameters as an optimization problem. Given an objective function of the form FITNESS*(Clusters C, Feedback F)* that indicates the quality of the set of clusters *C* in the light of the feedback *F*, the problem is to search for configuration parameters that yield clusters that maximize FITNESS. Rather than using a model of the problem to estimate the quality

Table 1. Entity resolution algorithm parameters.

Parameter	Description	Type	Optimization range
numKeys	Number of keys per tuple in HASH	Integer	$1 \leq numKeys \leq 9$
keyComponents	Number of values contributing to a key in HASH	Integer	$N/A - keyComponents = 1$
q	Size of q-gram in HASH	Integer	$N/A - q = 3$
H_i^1	Index into collection of first level hash functions	Integer	$1 \leq H_i^1 \leq 100$
H_i^2	Index into collection of second level hash functions	Integer	$1 \leq H_i^1 \leq 100$
K	Nearest neighbors returned by KNEARESTNEIGHBORS	Integer	$1 \leq k \leq 20$
similarityThreshold	Maximum distance between in KNEARESTNEIGHBORS candidate duplicates	Float	$0 \leq similarityThreshold \leq 1.0$
membershipThreshold	Voting threshold in MOSTLIKELYCLUSTER	Float	$0 \leq membershipThreshold \leq 1.0$
w_i	Attribute weights in DISTANCE, such that there is one weight per attribute	Float	$0 \leq w_i \leq 1.0$

of a candidate solution (as, for example, is standard practice for query optimization), the search takes place over the space of possible parameter settings, and the FITNESS is computed using clusters produced by running ENTITYRESOLUTION with candidate parameters over real data and feedback. Such an approach is followed because there is no known model for predicting the quality of the result of the entity resolution process given its parameters.

For entity resolution, we assume that feedback takes the form of annotations on pairs of records that are candidate duplicates, that confirm or refute that a pair of records are truly duplicates; such feedback is also obtained in [31,33]. We then require a fitness measure for the outcome of the entity resolution process, the clusters, that builds on this feedback. Our fitness function uses the fraction of the feedback that has been correctly clustered as an estimate of the fitness that results from the use of a set of configuration parameters. In deriving this fraction, we use the notation FC to denote counts derived from the feedback and the clustering, where F and C can each take the values M or U. The first character, F, represents the status of a pair of records in the Feedback: an M indicates that a pair of records match (i.e. they are duplicates); and a U indicates that a pair of records is unmatched (i.e. they are not duplicates). The second character, C represents the status of a pair of records in the clusters: an M indicates that a pair of records match (i.e. are duplicates); and a U indicates that a pair of records is unmatched (i.e. are not duplicates). Thus the notation:

- *MM* denotes the number of records that are matched in the feedback that are also matched in the clusters.
- *UU* denotes the number of records that are unmatched in the feedback that are also unmatched in the clusters.
- *MU* denotes the number of records that are matched in the feedback that are unmatched in the clusters.
- *UM* denotes the number of records that are unmatched in the feedback that are matched in the clusters.

The FITNESS of a clustering in the context of some feedback is then defined as:

$$\frac{MM + UU}{MM + MU + UU + UM}$$

Intuitively, this is the fraction of the feedback which is correctly represented in the clusters. We note that this notion differs from existing fitness measures for clusters, but that existing measures tend either: (i) to assume access to the ground truth (e.g. [14]), of which we have access to only a portion via feedback; or (ii) to relate to general properties of clusters such as inter- and intra-cluster distance (e.g. [23,27]) that may not be good indicators of how appropriately values have been assigned to clusters.

3.2   Evolutionary Search

This section describes how an evolutionary search, in particular a genetic algorithm, can be used to optimize the configuration parameters of the entity resolution technique from Sect. 2. We chose to employ a genetic algorithm because: (i) the search space is large, precluding the use of an exhaustive search; (ii) the search acts over a heterogeneous collection of configuration parameters, which are nevertheless easily accommodated in genetic algorithms; (iii) a genetic algorithm often takes less time to complete a task than other search methods [25], which is important in our context because of the high cost of clustering which is used to support fitness evaluation; and (iv) fitness evaluation can be easily parallelized in an evolutionary search.

We continue with some terminology [24]. An evolutionary search maintains an evolving *population* of interim solutions (individuals), which is subject to changes inspired by genetics. In our case, each member of the population consists of a collection of values for configuration parameters that control how entity resolution is carried out (see Table 1); the overall approach should be able to be applied to other entity resolution strategies using a population that captures their configuration parameters in place of those in Table 1). A *parent* is a member of a population that has been chosen to participate in the production of the next population (or generation) by way of *mutation* and *crossover* operators. A *mutation* introduces a random change to an individual member of the population, whereas a *crossover* combines features from two parents to produce a new member of the next population. A *fitness* function models requirements

---

Algorithm 4. GENETICSEARCH(Set<Tuple>Data, Feedback F)

---
1 $population \leftarrow$ initial collection of new random individuals
2 $fitness \leftarrow$ initial assignment of 0 for each individual
3 $bestIndividual \leftarrow null$
4 $bestFitness \leftarrow 0.0$
5 $n \leftarrow$ number of elite individuals to be retained
6 $t \leftarrow$ tournament size
7 $c \leftarrow$ crossover rate
8 $m \leftarrow$ mutation rate
9 $counter \leftarrow 0$
10 **while** $counter < generation\ size$ **do**
11   **for** $i \leftarrow 1$ **to** $populationsize$ **do**
12           $Clusters \leftarrow \text{CLUSTER}(Data, population[i])$
13           $fitness[i] \leftarrow \text{FITNESS}(Clusters, F)$
14   **if** $bestFitness < fitness[i]$ **then**
15               $bestIndividual \leftarrow population[i]$
16               $bestFitness \leftarrow fitness[i]$

17       $nextPopulation \leftarrow$ the n fittest individuals in $population$
18       $count \leftarrow 0$
19   **while** $count < populationsize/2$ **do**
20           $parent_a \leftarrow TournamentSelection(population, t)$
21           $parent_b \leftarrow TournamentSelection(population, t)$
22           $children \leftarrow Crossover(parent_a, parent_b, c)$
23           $nextPopulation = nextPopulation \cup Mutate(children, m)$
24           $count = count + 1$

25       $population = nextPopulation$
26       $counter = counter + 1$
27 Return $bestIndividual$

---

of the search problem, and informs the selection of parents, although to ensure diversity in the population not only the fittest solutions act as parents.

An evolutionary search starts from a set of random solutions that constitutes the initial population. The following steps are then repeated, as detailed in Algorithm 4, until a termination condition (e.g., the number of generations) is satisfied:

– **Fitness evaluation** assigns a fitness to each element in the population (lines 13–16). In our case, the fitness of an individual (i.e. a collection of values for configuration parameters) is obtained by generating a collection of clusters C using those configuration parameters, and then by using the definition of FITNESS from Sect. 3.1 to assess these clusters in the light of the user feedback F.
– **Elite capture** maintains a set of the best solutions found to date (line 17); the elites then have a role in the construction of future generations, and have the effect of focusing the search on promising solutions. This focusing can reduce diversity in the population, but is used here because the high cost of fitness evaluation means that we cannot afford too many generations.

- **Parent selection** chooses from the current generation those that should be used as parents for the next generation (lines 20–21). We apply tournament selection, in which the likelihood of an individual being a parent is proportional to its fitness.
- **Crossover** is a binary variation operator on two parents that brings together properties of different individuals (line 22). We use one-point crossover, which splits both parents at a randomly selected element and exchanges their tails. The crossover is applied with a probability c, called the *crossover rate*.
- **Mutation** is a unary variation operator that manipulates a single parent, and is used to maintain diversity in a population (line 23). Each of the values in our population is numeric, and we use *gaussian convolution* to mutate these values in such a way that the values in the children are likely to be similar to those of their parents, but periodically are substantially different [24]. The mutation is applied with a probability m, called the *mutation rate*.

4   Evaluation of Pay-as-you-go Clustering

This section describes the experiments carried out to evaluate the effectiveness of the parameter optimization strategy described in Sect. 3, given different amounts of user feedback. The purpose of the evaluation is to investigate the extent to which the strategy can improve on default configurations, and the amount of feedback required to obtain such improvements.

4.1   Experimental Setup

Datasets. We use three real datasets made available by University of Leipzig, for which the ground truth is provided with the dataset, that have been widely used in other studies of the performance of entity resolution techniques (e.g. [21, 22, 30, 31]):

- *DBLP-ACM*[2] is a data set containing bibliographic records from 2 data sets. There are a total of *12,051,595* pairs of records, of which *1083* represent the same entity. An example record is: <*672969, "An Effective Deductive Object-Oriented Database Through Language Integration", "Maria L. Barja, Norman W. Paton, Alvaro A. A. Fernandes, M. Howard Williams, Andrew Dinn", "Very Large Data Bases", 1994>*.
- *Abt-Buy*[3] is a data set containing *2173* product records from 2 data sets. There are a total of *2,359,878* pairs of records, of which *1097* represent the same entity. An example record is: <*6493, "Denon Stereo Tuner - TU1500RD", "Denon Stereo Tuner - TU1500RD/RDS Radio Data System/AM-FM 40 Station Random Memory/Rotary Tuning Knob/Dot Matrix FL Display/Optional Remote", $375.00>*.

---

[2] dbs.uni-leipzig.de/file/DBLP-ACM.zip.
[3] dbs.uni-leipzig.de/file/Abt-Buy.zip.

- *Amazon-Google*[4] is a data set containing *4598* product records from 2 data sets. There are a total of *10,527,166* pairs of records, of which *1,300* represent the same entity. An example record is: <http://www.google.com/base/feeds/ snippets/12244614697089679523, *"production prem cs3 mac upgrad,adobe cs3 production premium mac upgrade from production studio premium or standard, adobe software",805.99*>.

Techniques. We evaluate several different approaches to entity resolution, all of which generate clusters using the strategy described in Sect. 2. These approaches differ along three principal dimensions: (i) whether or not optimization takes place, as described in Sect. 3; (ii) when optimization does take place, whether only weights, or both weights and parameters participate in the search space; and (iii) whether or not the feedback is applied directly to tuple pairs in a way that overrides the score from the distance function used during clustering. In (iii), where a tuple pair is known from feedback to represent the same entity it can be given the minimum distance of *0*, and where a pair is known from feedback to represent different entities it can be given the maximum distance of *1*.

Table 2. Approaches compared.

Approach	Optimization	Parameters optimized	Score change
Baseline	No	Not applicable	No
NOSC	No	Not applicable	Yes
WOO	Yes	w_i in Table 1	No
WAPO	Yes	All of Table 1	No
POSC	Yes	All of Table 1	Yes

The approaches evaluated are as follows, where their key features are summarized in Table 2:

- *Baseline:* this method makes no use of feedback, and simply applies the entity resolution strategy from Sect. 2 directly. As such, the baseline has been subject to manual configuration. The weights w_i used by the DISTANCE function in Table 1 are problem-specific, and were set to values that give higher values to properties that seem likely to be of greater significance for matching, specifically: DBLP-ACM – (title: 0.8, authors: 0.4, venue: 0.6, year: 0.7); AbtBuy – (product_name: 0.6, description: 0.8, price: 0.2); and Amazon-Google – (product_name: 0.8, description: 0.4, manufacturer: 0.6, price: 0.2). The other parameter values in Table 1 were set as follows: $numKeys = 9$, $keyComponents = 1$, $q = 3$, H^1 and H^2 are chosen at random, $K = 10$,

---

[4] dbs.uni-leipzig.de/file/Amazon-GoogleProducts.zip.

similarityThreshold = 0.25, *membershipThreshold* = 0.5. Where specific values have been given, these were always based on positive experiences with those values in practice, so the baseline should be representative of what can realistically be achieved based on general experience.

– *No-optimization score change (NOSC):* this method applies the entity resolution strategy from Sect. 2 without parameter optimization, but with scores changed to reflect feedback that confirms or refutes that tuple pairs represent the same entity. This is to allow us to understand the effectiveness of a strategy in which the feedback is used to change scores but not to inform the configuration of the entity resolution process.

– *Weights-only optimization (WOO):* this method uses the optimization method from Sect. 3, but only searches over the weights used in the distance function (w_i in Table 1). This is to allow us to understand the relative impacts of optimization of the distance function and the optimization of the control parameters.

– *Weights-and-parameters optimization (WAPO):* this method uses the optimization method from Sect. 3, and searches over not only the weights but also the control parameters in Table 1.

– *Post-optimization score change (POSC):* this method uses the optimization method from Sect. 3, and searches over all the parameters in Table 1, but *after* the optimization scores are changed to reflect feedback that confirms or refutes that tuple pairs represent the same entity, and the dataset is reclustered with the parameters from the optimization and the changed scores[5]. This it to allow us to understand if optimization of the weights and parameters is sufficient to bring about appropriate clustering of the records on which feedback has been obtained.

We are unable to carry out a direct head-to-head comparison with related work, such as that discussed in Sect. 6, as there is no other proposal that covers parameter setting and distance function tuning, from blocking to clustering.

Scope of Optimization. Note that where the optimization is referred to as acting on the control parameters in Table 1, in fact for three of these parameters (*q, keyComponents* and *numKeys*), a sensitivity analysis identified that specific values (3, 9 and 1 respectively) always yielded the best results for our data sets, and thus these do not participate in the optimization.

---

[5] Note that in principle, it is also possible to have some form of *pre-optimization score change* strategy, which changes scores to reflect feedback that confirms or refutes that tuple pairs represent the same entity *before* optimization takes place. However, in practice, such an approach was found to be ineffective for optimizing configuration parameters because the score changes in themselves tend to be sufficient to ensure appropriate clustering of the feedback. Thus changes to configuration parameters tend to have little impact on fitness, and are not necessarily helpful for informing the clustering the records for which feedback has not been obtained.

Evolutionary Search. The optimization was implemented using the ECJ evolutionary computing system[6]. Following experimentation with different values, it was ascertained that using a *population size* of *50* and a generation size of *70* yielded results that were both stable and close in quality to those obtained with more searching; as time consuming entity resolution takes place during optimization, it is important that the search can converge on appropriate answers after a modest number of generations.

Feedback Generation. The experiments use feedback that is generated algorithmically from the ground truth. In essence, the feedback is a subset of the ground truth, which assigns an annotation *match* or *unmatch* to pairs of candidate duplicates. The approach makes few specific assumptions about how the feedback is obtained, although we do need both *match* and *unmatch feedback*; it is possible that targeted selection of feedback of the form discussed in Sect. 6 could produce improved results, but for now we use a straightforward approach to sample from the ground truth.

To generate *match* feedback, matching pairs are selected at random from the ground truth. Generating plausible *unmatch* feedback requires more care. In any data set, there are likely to be many more *unmatched* pairs than *matched* pairs, and in practice it makes little sense to ask users if a randomly selected pair of records match, as generally they will not match and there will be no reason to believe that they do. So, to generate *unmatch* feedback, the following steps take place: (i) generate the collection of *blocked pairs* by running the blocking algorithm described in Sect. 2 – *blocked pairs* should contain only pairs for which there is some evidence of syntactic similarity; and (ii) select pairs at random from the *blocked pairs* that have the annotation *unmatch* in the ground truth. In the experiments, we have identical amounts of *match* and *unmatch* feedback.

There is no fully accepted way of selecting data for feedback for entity resolution; indeed identifying the most appropriate data on which to obtain feedback is a research topic in its own right. The approach to feedback generation used in the experiment shares features with other work in the literature. For example, in investigating interaction with crowds for entity resolution, CrowdER [31] uses machine-based techniques to identify candidates, on which feedback is then obtained. This is analogous to our use of the blocking algorithm to identify candidates for *unmatch* feedback.

Evaluating Clusters. The following formula is used to compute the fitness of clustering results with respect to the ground truth:

$$\left(\frac{M_G M}{M_G M + M_G U} + \frac{U_G U}{U_G U + U_G M}\right) \times \frac{1}{2}$$

where:

---

[6] http://www.cs.gmu.edu/~eclab/projects/ecj/.

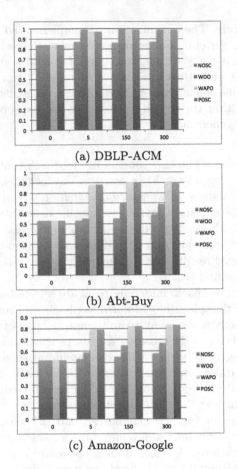

(a) DBLP-ACM

(b) Abt-Buy

(c) Amazon-Google

Fig. 1. Experiment 1: cluster quality for different levels of feedback for all 4 methods. The results for no feedback are the baseline for the experiment.

- $M_G M$ denotes the number of records that are matched in the ground truth that are also matched in the clusters.
- $U_G U$ denotes the number of records that are unmatched in the ground truth that are also unmatched in the clusters.
- $M_G U$ denotes the number of records that are matched in the ground truth that are unmatched in the clusters.
- $U_G M$ denotes the number of records that are unmatched in the ground truth that are matched in the clusters.

This calculates the fraction of correctly clustered record pairs over all record pairs, giving equal importance to the correct clustering of the matching and unmatching portions of the result. This approach is adopted in a context where there are generally many more unmatched pairs than matched pairs, to ensure that the counts of unmatched pairs do not swamp those for matched pairs.

The formula used for evaluating the clusters in the experiments is similar to that used for estimating the fitness of clusters in Sect. 3.1. The main difference is that for evaluating clusters we have access to the ground truth, whereas the fitness function used when creating the clusters in Sect. 3.1 only has access to the feedback, which is an approximation of the ground truth. The overall approach of computing fractions of correctly clustered data in the evaluation is standard (e.g. [14]).

4.2 Results

Experiment 1

Comparison of different approaches, for three different levels of feedback. The aim of this experiment is to understand the relative performance of the different methods from Table 2, so that subsequent experiments can investigate the most promising approaches in more detail. The experiment involves all three datasets (DBLP-ACM, AbtBuy and AmazonGoogle), and three different levels of feedback. Each item of feedback represents the annotation of a single candidate pair as *match* or *unmatch*. As the data sets have different sizes, as detailed in Sect. 4.1, the amounts of feedback experimented with range from annotations involving less than 1% of the records in each data set, up to individual annotations on around 6% of the records in DBLP-ACM and AmazonGoogle, and 14% of the records in AbtBuy.

The results are in Fig. 1(a) for DBLP-ACM, in Fig. 1(b) for AbtBuy and in Fig. 1(c) for AmazonGoogle. Where there is no feedback, this corresponds to the *Baseline* method described in Sect. 4.1. The following can be observed: (i) For NOSC, the feedback yields limited improvements in all cases – the local edits to scores are applied in the context of the default weights and parameter settings, and result in only small scale improvements to clusters. (ii) For WAPO and POSC, for all data sets, even a small amount of feedback yields a substantial improvement in performance. (iii) For the optimization based strategies, WOO, WAPO and POSC, the performance has leveled off by the time there are 150 items of feedback; other feedback amounts are considered in Experiment 2. (iv) The performance of WAPO is much better than for WOO in Abt-Buy and Amazon-Google, showing that there is benefit to be derived from including control parameters as well as weights in the optimization; several proposals from the literature focus on the optimization of comparison functions without also considering control parameters (e.g. [13,15]). (v) The performance of POSC is very similar to that of WAPO, showing that WAPO generally clusters the data items for which there is feedback correctly, without the additional evidence that comes from changed scores.

Experiment 2

Comparison of selected approaches for different amounts of feedback. The aim of this experiment is to understand the rate at which the quality of a clustering can be expected to improve as the amount of feedback collected grows. Thus this experiment is critical in terms of the cost-effectiveness of the pay-as-you-go approach. The experiment involves all three datasets (DBLP-ACM, AbtBuy, AmazonGoogle) and varying levels of feedback. The experiment focuses on NOSC and WAPO: NOSC as it represents a baseline in which the feedback is used but there is no optimization, and WAPO because it emerged as a strong proposal from Experiment 1.

The results are in Fig. 2(a) for DBLP-ACM, in Fig. 2(b) for AbtBuy and in Fig. 2(c) for AmazonGoogle. The following can be observed: (i) In WAPO, small amounts of feedback yield results that are close to the results obtained with much larger amounts of feedback; this is a positive result, as it suggests that the approach is cost-effective in terms of feedback collection. For AbtBuy, 5 items of feedback involves around 0.2% of the records, and for DBLP-ACM and AmazonGoogle, 5 items of feedback involves around 0.1% of the records. (ii) The highest quality measure obtained is always quite high, but varies between

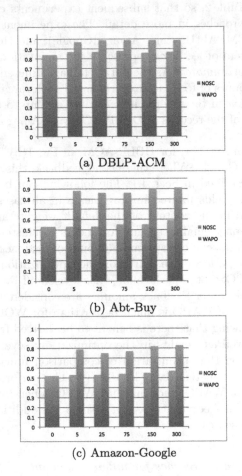

(a) DBLP-ACM

(b) Abt-Buy

(c) Amazon-Google

Fig. 2. Experiment 2: cluster quality for fine grained levels of feedback for NOSC and WAPO. The results for no feedback are the baseline for the experiment.

data sets. This is because in some data sets the syntactic evidence on similarity is stronger than in others; for example there are rarely inconsistencies in titles between the same paper in DBLP-ACM, but product descriptions for the same product are often significantly different in AbtBuy and AmazonGoogle. (iii) In AbtBuy and AmazonGoogle, for WAPO the result quality is slightly less good for 25 and 75 items of feedback than for 5 items of feedback. This is not especially surprising: (i) in all cases the feedback represents a small fraction of the ground truth, which means that the estimated fitness is subject to fluctuations based on the specific data for which feedback has been obtained; and (ii) the search for solutions is not exhaustive, and thus the search itself gives rise to variations in result quality.

5   Scaling the Approach

An issue with the approach in Sect. 3 is the high cost of evaluating the fitness function for the evolutionary search, which involves repeatedly applying the entity resolution technique from Sect. 2 on the data set for each candidate solution. Where there is a population size of p and g generations, this leads to $p \times g$ runs of the clustering algorithm. Although the search can readily be parallelized using platforms such as map/reduce [8] or Condor [29] so that the fitness of every element in the population at a generation is computed in parallel (indeed, our implementation runs using Condor on a campus grid), this can still be both resource intensive and lead to substantial elapsed times of broadly $g \times c$, were c is the cost of clustering. This section makes a proposal for reducing the cost of exploring the space of candidate solutions, and evaluates its effectiveness both in terms of runtime and cluster quality.

5.1   Clustering on Pruned Data Sets

With a view to reducing response times and resource usage, here we describe an approach to reducing c by clustering only the portion of the dataset on which feedback has been obtained. Recall from Sect. 3.1 that the fitness of a clustering is estimated from the fitness of the feedback (which represents the available subset of the ground truth). Thus although changes to the weights and parameters of the clustering apply to every item in the data set, the estimated fitness depends only on the clustering of the pairs for which we have feedback. Thus, with a view to reducing the computational cost that results from clustering records about which we have no feedback, clustering is run over a pruned dataset that consists of the records on which we have feedback, which is defined as follows:

$$prunedDataset = \{r | r \in dataset, hasFeedback(r)\}$$

where *dataset* is the collection of records to be clustered, and *hasFeedback* is true for a record if there is any feedback on that record.

Thus when pruning is used, during optimization the fitness of a candidate configuration is an estimate in two senses: (i) the fitness is based only on the feedback, and not on the ground truth, which is not available in practice; and (ii) the fitness is calculated over clusters that involve only a subset of the complete dataset, in contrast with Sect. 4.2. In reporting the results of the experiments, cluster quality is reported in terms of the ground truth for a clustering of the complete dataset, where that clustering was obtained using the configuration parameters obtained when optimizing using the pruned dataset.

Table 3 indicates the number of records in the pruned data sets for different amounts of feedback. Where the feedback amounts are small, the number of records is typically twice the number of items of feedback, as each item of feedback involves the relationship between two records. As the amount of feedback grows, there is a growing likelihood that a new item of feedback will involve at least one record for which there is already some feedback, and hence the number of records becomes less than twice the amount of feedback.

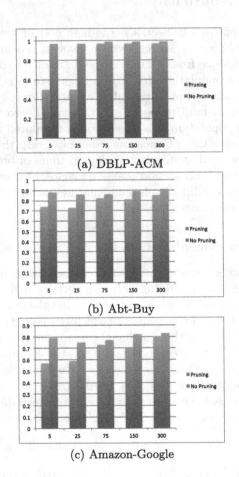

(a) DBLP-ACM

(b) Abt-Buy

(c) Amazon-Google

Fig. 3. Experiment 3: cluster quality for different levels of feedback using pruned and non-pruned data sets for WAPO.

Table 3. Pruned data set size for different feedback amounts.

Dataset	Total number of records	Feedback amount				
		5	25	75	150	300
DBLP-ACM	4910	10	50	150	292	572
AbtBuy	2173	10	50	146	278	535
Amazon Google	4589	10	50	149	295	578

5.2   Evaluation of Pruning

This section empirically evaluates the effectiveness of the pruning approach in terms of runtime performance and the quality of the clusters produced. The purpose of the evaluation is to investigate the extent to which pruning can improve on the runtime required for a complete clustering, and the extent to which this reduced runtime is accompanied by reduced cluster quality.

Although the fitness function is applied only to records for which there is feedback, the pruning of records for which there is no feedback can be expected to impact on the clustering of the records for which there is feedback, for example because such records now have different neighbors. As such, it is important to evaluate the results of the pay-as-you-go entity resolution strategy to ascertain: (i) the extent to which effective weights and parameters are obtained when optimization takes place over the pruned data sets; and (ii) the performance improvement in terms of clustering times.

Experiment 3

Comparison of cluster quality for optimization using pruned data sets. The aim of this experiment is to understand the extent to which the quality of clusters produced for a given level of feedback is affected by optimization over pruned data sets. Thus this experiment is important in terms of the scalability of approach. The experiment involves all three datasets (DBLP-ACM, AbtBuy and Amazon-Google) with varying levels of feedback. The experiment focuses on WAPO as it has been shown to perform well, and because it depends for its performance on the effectiveness of the optimization.

The results are reported in Fig. 3. The following can be observed: (i) In all cases, where there are 75 or more items of feedback, representing a single item of feedback on 2% to 4% of the records, the results obtained using pruned data sets significantly improve on the default parameters, for which the results were reported in Experiment 1. (ii) In all cases, there is a gap between the results with and without pruning, which narrows as more feedback is obtained. The gap would be expected to narrow, as increased amounts of feedback leads to larger and more representative data sets being used in the assessment of the fitness of candidate solutions. (iii) For the smaller amounts of feedback, specifically 5 and 25 items, the gap between the pruned and non-pruned cases can be large, and the results using pruning are worse than the default parameters in DBLP-ACM.

This is explained by the clustering taking place on tiny data sets, which are unrepresentative.

This experiment illustrates that there is an interesting trade-off between the two kinds of payment in pay-as-you-go entity resolution. Payment in the form of user feedback provides a more rapid return on investment when fitness is evaluated by clustering the complete data set. However, clustering the complete data set within the search is computationally expensive. By contrast, with the pruned data sets the cost of clustering can be significantly reduced, but good results can only be obtained where there is more feedback.

Experiment 4
Comparison of clustering times between original and pruned data sets. The aim of this experiment is to understand the impact of pruning on the runtime of clustering. The experiment uses all three data sets, and focuses on WAPO as it has been shown to be an effective strategy.

In practice, clustering times on complete data sets can vary quite significantly between runs (up to about a factor of 3), as different configurations lead to different collections of neighbors, etc. As such, to give a summary of the improvements that can be expected, Table 4 reports the average speedup obtained, averaged over 5 runs, using the pruned data set in place of the complete data set.

Table 4. Average runtimes (seconds) and speedup obtained using pruned data set with 300 items of feedback.

Dataset	Runtime: full	Runtime: pruned	Speedup
DBLP-ACM	300	5	60
AbtBuy	378	14	27
Amazon Google	9360	300	31

The results show that clustering with the pruned data set is several orders of magnitude faster than with the complete data set; even more impressive results can be obtained for more intensive pruning. This shows that pruning can be used with large data sets to very substantially reduce the cost of fitness function evaluation.

6 Related Work

This section discusses some of the most relevant related work, focusing on *learning/optimization for entity resolution* and *crowdsourcing for entity resolution*, both of which seek to improve the results of the entity resolution process on the basis of feedback or training data. In discussing these areas, although there are proposals that follow a pay-as-you-go approach, there is no other proposal that covers parameter setting and distance function tuning, from blocking to clustering.

We do not review the wider literature on entity resolution, where there are existing surveys on the topic as a whole [11,20], on blocking [4] and on clustering [14].

In relation to *learning/optimization for entity resolution*, there has been work to inform both blocking strategies and more detailed comparison rules. Although not specifically following a pay-as-you-go approach, several relevant proposals have been made that learn or tune blocking schemes, for example using training data to inform the selection of properties that participate in blocking functions (such as [3,26]), or by tuning similarity thresholds (e.g. [2]). More detailed rules for comparing values have also been learned using genetic programming [6], for example for identifying links between linked open data resources [15]. In addition, as entity resolution potentially acts over huge data sets, research has been carried out into the use of active learning for obtaining the most suitable training data [1,16]. Such research complements the work in this paper; in this paper the focus is on co-optimizing blocking, comparison and clustering, and insights from work on each of these individual stages can help to identify opportunities for their combined optimization.

In our work, in order to evaluate the fitness of a candidate set of parameters, the full entity resolution process needs to be re-run for each candidate. Such an approach, known as closed-loop optimization, has been widely used in other domains to automatically set configuration parameters for physical systems; for example, Knowles [19] describes the use of evolutionary search techniques for tasks as diverse as optimizing instrument setup in analytical biochemistry, and improving chocolate production. This practice of searching for suitable parameters by running the actual system has also been employed for computing systems. For example, in iTuned [10], experiments are generated that investigate the effect of different system parameters on overall performance, in a context, like ours, where the development of an analytical cost model is not obviously a practical proposition. Furthermore, in reinforcement learning [18], learning is intrinsically closely associated with an ongoing process, with the normal behaviour of a system interleaved with learning steps that follow a trial-and-error approach. For entity resolution, de Freitas *et al.* [7] combine active learning with reinforcement learning, where the latter is used to evaluate the confidence of different committee members that are making detailed comparison decisions.

In relation to *crowdsourcing for entity resolution*, there is work on *identifying the most suitable data on which to crowdsource feedback* and on *applying the feedback to inform entity resolution decisions*.

For *identifying the most suitable data on which to crowdsource feedback*: CrowdER [31], focuses on the grouping of candidate entity pairs into tasks with a view to obtaining the required information while minimizing the number of crowd tasks, and thus the expense; and several proposals have investigated the identification of the pairs of records that are likely to be most valuable for refining decisions relating to entity resolution (e.g. [16,30,32,33]). For example, in Isele *et al.* [16], feedback is sought using active learning on the record pairs on which candidate detailed comparison rules disagree the most.

Such research complements the work in this paper, which could be used alongside techniques for efficient collection of feedback to improve overall return on investment.

For *applying the feedback to inform entity resolution decisions*, several proposals consult the crowd for specific purposes. ZenCrowd [9] identifies pairs of instances in linked data using two levels of blocking to identify candidate pairs for confirmation by the crowd. A probabilistic *factor graph* accumulates evidence from different sources, from which a probability is derived that a candidate pair is correct. The Silk Link Discovery Workbench [16] uses the crowd to judge whether candidate pairs are duplicates, and refines the collection of detailed comparison rules on the basis of this feedback. Perhaps the work that is closest to ours in terms of ethos is Corleone [13], which seeks to provide *hands-off crowdsourcing* for entity resolution, whereby the whole entity resolution process is automated, obtaining input from the crowd as needed. To do this, Corleone uses crowdsourcing to learn blocking and refinement rules in turn, and also addresses issues such as when to terminate a crowdsourcing activity. Our work is similar, in including several entity resolution phases, but is somewhat broader in scope in that it also optimizes related system parameters and considers clustering as well as pairwise matching. In addition, the approaches are significantly different; Corleone tackles blocking and matching in sequence, whereas in this paper, blocking, the distance function and other system thresholds and parameters are optimized at the same time. The price paid to support this co-optimization is the need to run blocking and clustering repeatedly within the evolutionary search, although we have shown that the costs associated with this can be reduced using pruning. In addition, by co-optimizing system parameters, our methodology reduces the risk that important but non-obvious interactions between parameters are overlooked during the application of the pay-as-you-go methodology.

Note that the term *pay-as-you-go* has been applied with different meanings in relation to entity resolution. In this paper, as in the above literature on crowdsourcing, the payment takes the form of feedback on the results of an entity resolution process, whereas in [34] the payment is in the form of computational resource usage.

7   Conclusions

We now revisit the claimed contributions of the paper from the introduction:

1. *A generic approach to configuration of parameters for entity resolution that uses feedback on the correctness (or otherwise) of candidate duplicates.* We have described an approach that uses closed loop optimization to evaluate the effectiveness of alternative configuration parameters, simultaneously optimizing all stages of the entity resolution process; this is a potentially important feature of the approach, as there may be subtle relationships between parameters across the process.

2. *A description of the application of that approach to produce a pay-as-you-go self-optimizing entity resolution system that uses an evolutionary search*

over the space of parameter values. We have applied the approach from (1) to automate the optimization of the parameters for the state-of-the-art approach described in Costa *et al.* [5]; the optimized parameters include weights within the distance function, indexing parameters and similarity thresholds.

3. *An evaluation of the resulting platform with real world data sets, which shows substantial improvements in cluster quality.* The evaluation shows not only that the technique can be effective, providing significantly improved clustering compared with default parameters, but also that effective results can be obtained with surprisingly modest amounts of feedback (a single item of feedback on less than 1% of the records in each of the experimental data sets). The experiments also showed that optimizing the control parameters and distance function together was more effective than optimizing only the distance function, a heretofore popular strategy.

4. *As the proposal can be considered to be computationally expensive, we describe and evaluate an approach that seeks to retain the results from (3) while also scaling to large data sets.* A modification of the approach is described in which the entity resolution process, rather than testing the fitness of candidate configurations over the complete data set, instead evaluates fitness over the (typically much smaller) portion of the data set for which feedback has been obtained. The experimental evaluation shows that more feedback is required to obtain stable and effective results than when fitness is evaluated over the complete dataset during the search. However, even when using the pruned data set, significant improvements over the default parameters have been obtained with feedback on a small percentage of the data (a single item of feedback on less than 4% of the records in each of the experimental data sets). Pruning on such amounts of feedback can reduce runtime costs of clustering by several orders of magnitude.

There are several possible areas for future work. Although positive results have been demonstrated with small amounts of feedback, it would be interesting to ascertain if active learning [28] could focus feedback collection on values that are still more effective, in particular in the case of the pruning strategy. In the application of the approach, the distance function is fairly straightforward; it would be interesting to investigate the co-optimization of richer distance functions (e.g. as in [6]) with other configuration parameters. Although the overall closed-loop optimization approach is, in principle, applicable for configuration of different entity-resolution algorithms, it would be interesting to demonstrate this in practice, and to ascertain the extent to which different parameters in different settings can be tuned effectively in this way. Furthermore, it would be interesting to investigate the use of optimization techniques that are specifically targeted at problems with expensive black-box fitness functions [17].

Acknowledgement. Research on data integration at Manchester is supported by the EPSRC VADA Programme Grant, whose support we are pleased to acknowledge.

References

1. Arasu, A., Götz, M., Kaushik, R.: On active learning of record matching packages. In: Proceedings of the ACM SIGMOD International Conference on Management of Data, SIGMOD 2010, Indianapolis, Indiana, USA, 6–10 June 2010, pp. 783–794 (2010)
2. Bianco, G.D., Galante, R., Heuser, C.A., Gonçalves, M.A.: Tuning large scale deduplication with reduced effort. In: SSDBM, p. 18 (2013)
3. Bilenko, M., Kamath, B., Mooney, R.: Adaptive blocking: learning to scale up record linkage. In: Sixth International Conference on Data Mining (ICDM 2006), pp. 87–96 (2006)
4. Christen, P.: A survey of indexing techniques for scalable record linkage and deduplication. IEEE Trans. Knowl. Data Eng. 24(9), 1537–1555 (2012)
5. Costa, G., Manco, G., Ortale, R.: An incremental clustering scheme for data deduplication. Data Mining Knowl. Disc. 20(1), 152–187 (2010)
6. de Carvalho, M., Laender, A., Goncalves, M., Da Silva, A.: A genetic programming approach to record deduplication. IEEE Trans. Knowl. Data Eng. 24(3), 399–412 (2012)
7. de Freitas, J., Pappa, G.L., da Silva, A.S., Gonçalves, M.A., de Moura, E.S., Veloso, A., Laender, A.H.F., de Carvalho, M.G.: Active learning genetic programming for record deduplication. In: Proceedings of the IEEE Congress on Evolutionary Computation (CEC 2010), Barcelona, Spain, 18–23, pp. 1–8, July 2010
8. Dean, J., Ghemawat, S.: Mapreduce: simplified data processing on large clusters. Commun. ACM 51(1), 107–113 (2008)
9. Demartini, G., Difallah, D.E., Cudré-Mauroux, P.: Large-scale linked data integration using probabilistic reasoning and crowdsourcing. VLDB J. 22(5), 665–687 (2013)
10. Duan, S., Thummala, V., Babu, S.: Tuning database configuration parameters with ituned. PVLDB 2(1), 1246–1257 (2009)
11. Elmagarmid, A., Ipeirotis, P., Verykios, V.: Duplicate record detection: a survey. IEEE Trans. Knowl. Data Eng. 19(1), 1–16 (2007)
12. Goiser, K., Christen, P.: Towards automated record linkage. In: Proceedings of the Fifth Australasian Conference on Data Mining and Analytics (AusDM 2006), vol. 61, pp. 23–31, Darlinghurst, Australia. Australian Computer Society Inc. (2006)
13. Gokhale, C., Das, S., Doan, A., Naughton, J.F., Rampalli, N., Shavlik, J.W., Zhu, X.: Corleone: hands-off crowdsourcing for entity matching. In: SIGMOD Conference, pp. 601–612 (2014)
14. Hassanzadeh, O., Chiang, F., Lee, H.C., Miller, R.J.: Framework for evaluating clustering algorithms in duplicate detection. Proc. VLDB Endow. 2(1), 1282–1293 (2009)
15. Isele, R., Bizer, C.: Learning expressive linkage rules using genetic programming. PVLDB 5(11), 1638–1649 (2012)
16. Isele, R., Bizer, C.: Active learning of expressive linkage rules using genetic programming. J. Web Sem. 23, 2–15 (2013)
17. Jones, D.R., Schonlau, M., Welch, W.J.: Efficient global optimization of expensive black-box functions. J. Global Optim. 13(4), 455–492 (1998)
18. Kaelbling, L.P., Littman, M.L., Moore, A.W.: Reinforcement learning: a survey. J. Artif. Intell. Res. (JAIR) 4, 237–285 (1996)
19. Knowles, J.D.: Closed-loop evolutionary multiobjective optimization. IEEE Comp. Int. Mag. 4(3), 77–91 (2009)

20. Kopcke, H., Rahm, E.: Frameworks for entity matching: a comparison. Data Knowl. Eng. **69**(2), 197–210 (2010)
21. Köpcke, H., Thor, A., Rahm, E.: Evaluation of entity resolution approaches on real-world match problems. PVLDB **3**(1), 484–493 (2010)
22. Lee, S., Lee, J., Hwang, S.-W.: Scalable entity matching computation with materialization. In: 20th ACM International Conference on Information and Knowledge Management (CIKM 2011), pp. 2353–2356. ACM (2011)
23. Legány, C., Juhász, S., Babos, A.: Cluster validity measurement techniques. In: 5th WSEAS International Conference on Artificial Intelligence, Knowledge Engineering and Data Bases (AIKED 2006), pp. 388–393. World Scientific (2006)
24. Luke, S.: Essentials of Metaheuristics. Lulu, Raleigh (2013)
25. Michalewicz, Z., Fogel, D.B.: How to Solve It: Modern Heuristics. Springer, Heidelberg (2004)
26. Michelson, M., Knoblock, C.A.: Learning blocking schemes for record linkage. In: Proceedings of the 21st National Conference on Artificial Intelligence (AAAI 2006), vol. 1, pp. 440–445. AAAI Press (2006)
27. Rendón, E., Abundez, I.M., Gutierrez, C., Zagal, S.D., Arizmendi, A., Quiroz, E.M., Arzate, H.E.: A comparison of internal and external cluster validation indexes. In: 2011 American Conference on Applied Mathematics and the 5th WSEAS International Conference on Computer Engineering and Applications (AMERICAN-MATH 2011/CEA 2011), pp. 158–163. World Scientific (2011)
28. Settles, B.: Active learning. Synth. Lect. Artif. Intell. Mach. Learn. **6**(1), 1–114 (2012). Morgan & Claypool Publishers
29. Thain, D., Tannenbaum, T., Livny, M.: Distributed computing in practice: the condor experience. Concurrency Pract. Exp. **17**(2–4), 323–356 (2005)
30. Vesdapunt, N., Bellare, K., Dalvi, N.: Crowdsourcing algorithms for entity resolution. Proc. VLDB Endow. **7**(12), 1071–1082 (2014)
31. Wang, J., Kraska, T., Franklin, M.J., Feng, J.: Crowder: crowdsourcing entity resolution. Proc. VLDB Endow. **5**(11), 1483–1494 (2012)
32. Wang, S., Xiao, X., Lee, C.: Crowd-based deduplication: an adaptive approach. In: Proceedings of the 2015 ACM SIGMOD International Conference on Management of Data, Melbourne, Victoria, Australia, May 31 - June 4 2015, pp. 1263–1277 (2015)
33. Whang, S.E., Lofgren, P., Garcia-Molina, H.: Question selection for crowd entity resolution. PVLDB **6**(6), 349–360 (2013)
34. Whang, S.E., Marmaros, D., Garcia-Molina, H.: Pay-as-you-go entity resolution. IEEE Trans. Knowl. Data Eng. **25**(5), 1111–1124 (2013)

A Unified View of Data-Intensive Flows
in Business Intelligence Systems: A Survey

Petar Jovanovic[✉], Oscar Romero, and Alberto Abelló

Universitat Politècnica de Catalunya, BarcelonaTech, Barcelona, Spain
{petar,oromero,aabello}@essi.upc.edu

Abstract. Data-intensive flows are central processes in today's business
intelligence (BI) systems, deploying different technologies to deliver data,
from a multitude of data sources, in user-preferred and analysis-ready formats. To meet complex requirements of next generation BI systems, we
often need an effective combination of the traditionally batched extract-
transform-load (ETL) processes that populate a data warehouse (DW)
from integrated data sources, and more real-time and operational data
flows that integrate source data at runtime. Both academia and industry thus must have a clear understanding of the foundations of data-
intensive flows and the challenges of moving towards next generation
BI environments. In this paper we present a survey of today's research
on data-intensive flows and the related fundamental fields of database
theory. The study is based on a proposed set of dimensions describing
the important challenges of data-intensive flows in the next generation
BI setting. As a result of this survey, we envision an architecture of a
system for managing the lifecycle of data-intensive flows. The results
further provide a comprehensive understanding of data-intensive flows,
recognizing challenges that still are to be addressed, and how the current
solutions can be applied for addressing these challenges.

Keywords: Business intelligence · Data-intensive flows · Workflow
management · Data warehousing

1 Introduction

Data-intensive flows are critical processes in today's business intelligence (BI)
applications with the common goal of delivering the data in user-preferred and
analysis-ready formats, from a multitude of data sources. In general, a data-
intensive flow starts by extracting data from individual data sources, cleans
and conforms extracted data to satisfy certain quality standards and business
requirements, and finally brings the data to end users.

In practice, the most prominent solution for the integration and storage of
heterogeneous data, thoroughly studied in the past twenty years, is data warehousing (DW). A DW system assumes a unified database, modeled to support
analytical needs of business users. Traditionally, the back stage of a DW system comprises a data-intensive flow known as the extract-transform-load (ETL)

A. Hameurlain et al. (Eds.): TLDKS XXIX, LNCS 10120, pp. 66–107, 2016.
DOI: 10.1007/978-3-662-54037-4_3

process responsible of orchestrating the flow of data from data sources towards a DW. ETL is typically a batch process, scheduled to periodically (e.g., monthly, daily, or hourly) load the target data stores with fresh source data. In such a scenario, limited number of business users (i.e., executives and managers) are expected to query and analyze the data loaded in the latest run of an ETL process, for making strategic and often long-term decisions.

However, highly dynamic enterprise environments have introduced some important challenges into the traditional DW scenario.

- Up-to-date information is needed in near real-time (i.e., right-time [39]) to make prompt and accurate decisions.
- Systems must provide the platform for efficiently combining in-house data with various external data sources to enable context-aware analysis.
- Systems must be able to efficiently support new, unanticipated needs of broader set of business users at runtime (i.e., on-the-fly).

These challenges have induced an important shift from traditional business intelligence (BI) systems and opened a new direction of research and practices. The *next generation BI* setting goes by various names: *operational BI* (e.g., [15,22]), *live BI* (e.g., [17]), *collaborative BI* (e.g., [6]), *self-service BI* (e.g., [1]), *situational BI* (e.g., [63]). While these works look at the problem from different perspectives, in general, they all aim at enabling the broader spectrum of business users to access a plethora of heterogeneous sources (not all being under the control of the user's organization and known in advance), and to extract, transform and combine these data, in order to make right-time decisions. Consequently, here, we generalize these settings and use the common term *next generation BI*, while for the old (DW-based) BI setting we use the term *traditional BI*. An interesting characterization of the *next generation BI* setting is given by Eckerson [22]: "...*operational BI requires a just-in-time information delivery system that provides the right information to the right people at the right time so they can make a positive impact on business outcomes.*"

Obviously, in such a scenario periodically scheduled batch loadings from pre-selected data stores have become unrealistic, since fresh data are required in near real-time for different business users, whose information needs may not be known in advance. In fact, effectively integrating the traditional, batched decision making processes, with on-the-fly data-intensive flows in *next generation BI* systems is discussed to be important to satisfy analytic needs of today's business users (e.g., [2,17,40]). For example, a part of in-house company's sales data, periodically loaded to a DW by an ETL, can be at runtime crossed with the Web data to make context-aware analysis of business decisions. To build such demanding systems, one must first have a clear understanding of the foundation of different data-intensive flow scenarios and the challenges they bring.

From a theoretical perspective, handling data heterogeneity has been separately studied in two different settings, namely *data-integration* and *data exchange*. Data integration has studied the problem of providing a user with a unified virtual view over data in terms of a global schema [59]. User queries over the global schema, are then answered by reformulating them on-the-fly in terms of data sources. On the other side, data exchange has studied the problem of materializing an instance of data at the target that reflects the source data as accurately as possible and can be queried by the user, without going back to the original data sources [27].

A recent survey of ETL technologies [96] has pointed out that the data exchange problem is conceptually close to what we traditionally assume by an ETL process. Intuitively, in an ETL process, we also create an instance at the target, by means of more complex data transformations (e.g., *aggregation, filtering, format conversions, deduplication*). However, the trends of moving towards the *next generation BI* settings have brought back some challenges initially studied in the field of data integration, i.e., requiring that the user queries should be answered by extracting and integrating source data at runtime. Moreover, the *next generation BI* settings brought additional challenges into the field of data-intensive flows (e.g., low data latency, context-awareness).

Right-time decision making processes demand close to zero latency for data-intensive flows. Hence the automated optimization of these complex flows is a must [17,44], not only for performance, but also for other quality metrics, like fault-tolerance, recoverability, etc. [84]. Considering the increasing complexity of data transformations (e.g., machine learning, natural language processing) and the variety of possible execution engines, the optimization of data-intensive flows is one of the major challenges for *next generation BI* systems.

Even though the above fields have been studied individually in the past, the literature still lacks a unified view of data-intensive flows. In this paper, we aim at studying the characteristics of data-intensive flows in *next generation BI* systems. We focus on analyzing the main challenges in the three main stages when executing data-intensive flows, i.e., (1) *data extraction*, (2) *data transformation*, and (3) *data delivery*. Having low data latency as an important requirement of data-intensive flows in the *next generation BI* setting, we additionally analyze the main aspects of *data flow optimization*. We analyzed these four areas inside the two main scenarios for data-intensive flows: (a) periodically executed, batched processes that materialize and load data at the target data store for future analysis (*extract-transform-load - ETL*), and (b) on-the-fly, instantaneous data flows executed on demand upon end-users' query (*extract-transform-operate - ETO*).

In addition, we identify that a well-known BigData challenge, namely, one of the so called 3 V's [45] (i.e., massive *volumes* of data), is an important one also for the design and even more deployment of data-intensive flows in next-generation BI systems. However, the approaches that deal with such challenge represent a separate and rather extensive field of research, which is out of scope of this study. We thus refer an interested reader to [13] for more detailed overview of the approaches dealing with the BigData challenges.

As the first result, we identify the main characteristics of data-intensive flows, focusing on those that best describe the challenges of moving towards the *next generation BI* setting. Then, in terms of these characteristics we classify the approaches, both from the foundational works and more recent literature, tackling these characteristics at different levels. On the one hand, the results provide us with a clear understanding of the foundations of data-intensive flows, while on the other hand, identified characteristics help us defining the main challenges of moving towards the *next generation BI* setting.

Finally, as the main outcome of this study, we outlined the envisioned architecture of *next generation BI* systems, focusing on managing the complete lifecycle of data-intensive flows.

Contributions. In particular, our main contributions are as follows.

- We analyzed current approaches, scrutinizing the main aspects of data-intensive flows in today's BI environments.
- We define the main characteristics of data-intensive flows, focusing on those that best describe the challenges of a shift towards the *next generation BI*.
- In terms of the dimensions defined from these characteristics, we analyze both the foundational work of database theory, and recent approaches for data-intensive flows, at different levels of these dimensions.
- Resulting from this study, we envision an architecture for managing the complexity of data-intensive flows in the *next generation BI* setting.
- Finally, we indicate the remaining challenges for data-intensive flows, which require further attention from both academia and industry.

Outline. In Sect. 2, we first introduce an example scenario used to support our discussions throughout this paper. We then in Sect. 3, describe the methodology used in our study, and outline the main study setting. Next, in Sect. 4 we discuss the process of defining the dimensions that are further used for studying data-intensive flows. In Sects. 5–8, we analyze different approaches from data-intensive flows inside the previously defined dimensions. In Sect. 9 we provide the overall discussion and introduce an envisioned architecture of a system for managing data-intensive flows in the *next generation BI* setting, while in Sect. 10, we conclude the paper.

2   Example Scenario

We first introduce an example scenario to support discussions throughout this paper and to motivate our study. Our example scenario is motivated by the data model introduced for the big data benchmark (a.k.a. BigBench) in [33], which extends the TPC-DS benchmark[1] for the context of big data analytics. Notice that we adapted their scenario to make the examples more intuitive and suitable to our discussions. In particular, besides the typical operations found in relational database systems, i.e., *Join*, Selection (*Filter*), Aggregation (*Aggr.*), *Sort*, and Distinct (*Remove Duplicates*), in the following example scenarios, we also introduce more complex operations typically found in today's data-intensive

---

[1] http://www.tpc.org/tpcds/spec/tpcds_1.1.0.pdf (last accessed 4/4/2014).

flows; that is, (1) User Defined Functions (*UDF*) that may implement either simple arithmetic expressions or complex, typically black-box operations, (2) *Match* that implements more relaxed join or lookup semantics (e.g., using approximate string matching), and (3) *Sentiment Analysis* that typically applies natural language processing techniques for extracting subjective (opinion) information from the Web sources (e.g., forums, social networks).

In general, we consider a simple case of a retail company that has different databases that support its daily operational processes. These databases cover the information about different items (i.e., products) offered for the sale, company's customers, their orders, shipping information, etc. Periodically, the company launches campaigns and puts some product on a promotion.

Scenario 1. Originally, the company has used a *traditional BI* system with a centralized DW that is loaded by means of different ETL flows (one of which is conceptually depicted in Fig. 1). Users in this scenario are typically upper management executives that analyze the enterprise-wide data (e.g., `items` and their `sales`) to make decisions for making strategic actions (e.g., launching promotional campaigns).

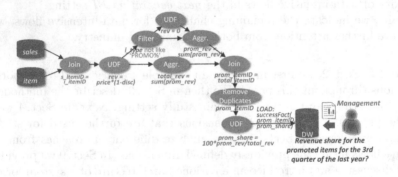

Fig. 1. Example 1.1: ETL to analyze revenue share from a promotion

Example 1.1. In the specific example in Fig. 1, the ETL flow is periodically executed to load the DW with information about the percentage of the revenue share made from the items that were on the promotion. Quarterly, the management analyzes how the previous business decisions on promotional campaigns affected the revenue.

Scenario 2. While the above setting has served the company well in having a periodical feedback about the previous strategic decisions, today's dynamic markets require more prompt reaction to the potentially occurring problems (e.g., hourly or daily). The company thus noticed that instead of waiting for the sales data to analyze the success of the promotional campaign, they can potentially benefit from the opinions that customer may leave about the campaign and product items, in the form of `reviews` over the Web (e.g., social networks) and react faster to improve the potential revenue. Moreover, the company also noticed that such an analysis should be decentralized to the regional and local representatives and available to a broader set of users involved in the business process. As fast decisions are needed, the users must be able to make them at right time (i.e., *"...before a problem escalates into a crisis or a fleeting opportunity disappears..."* [22]). We consider the two following business requirements posed on-the-fly, and the two data-intensive flows that answer them, conceptually depicted in Figs. 2 and 3.

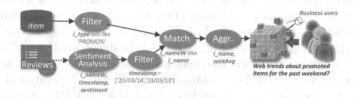

Fig. 2. Example 2.1: ETO to predict the success of a promotion

Example 2.1. In the former example (Fig. 2), a regional manager decides to analyze the potential success of the promotional campaign launched on Friday, after the first weekend, by inspecting the sentiment (i.e., opinions) of the real and potential customers about the product `items` that are included in the promotion.

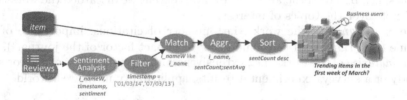

Fig. 3. Example 2.2: ETO to analyze the trends for launching a new promotion

Example 2.2. The latter example (Fig. 3), on the other hand, analyzes the currently trending product `items` and user opinions about them for deciding which items to include in the next promotional campaign.

In both cases, business users interactively make on-the-fly and context-aware analysis, in order to quickly react and improve their business decisions.

3   Methodology

We further introduce the methodology used for studying data-intensive flows and outline the resulting study setting. We start by introducing the process of selecting the literature to be included in this study. The study further includes three consecutive phases, which we explain in more detail in the following subsections.

3.1   Selection Process

The literature exploration started with the keyword search of the relevant works inside the popular research databases (i.e., Scopus[2] and Google Scholar[3]). In Phase I, we focused on the keywords for finding seminal works in the DW and ETL field (i.e., "data warehousing", "ETL", "business intelligence"), as well as the most relevant works on the *next generation BI* (i.e., "next generation business intelligence", "BI 2.0", "data-intensive flows", "operational business intelligence"). While in the case of traditional DW and ETL approaches we encountered and targeted the most influential books from the field (e.g., [46,53]) and some extensive surveys (e.g., [96]), in the case of the next generation BI approaches, we mostly selected surveys or visionary papers on the topic (e.g., [1, 2,15,17,40]). Furthermore, the following phases included keyword search based on the terminology found in the created study outline (see Fig. 4), as well as the identified dimensions (see Fig. 5).

Rather than being extensive in covering all approaches, we used the following criteria for prioritizing and selecting a representative initial set of approaches that we studied.

- The *relevance* of the works to the field of data-intensive flows and the related topics, i.e., based on the abstract/preface content we discarded the works that did not cover the topics of interest.
- The *importance* of the works, i.e., number of citations, importance of the venue (e.g., ranking of the conference[4] or impact factor of the journal[5]).
- The *maturity* of the works, i.e., extensiveness and completeness of a theoretical study or a survey, experimental results, applicability in the real world.

Furthermore, we also followed the snowballing technique and included, previously not found, but relevant approaches referenced from the initial ones.

3.2   Phase I (Outlining the Study Setting)

First phase included the review of the seminal works on traditional data-intensive flows, ETL, and data warehousing in general; as well as the relevant works discussing the *next generation BI* systems and their main challenges on moving toward (near) real-time data analysis.

---

[2] https://www.scopus.com/.
[3] https://scholar.google.com.
[4] CORE conference ranking: http://portal.core.edu.au/conf-ranks/.
[5] Thomas Reuters Impact Factor: http://wokinfo.com/essays/impact-factor/.

As a result, we outline the setting for studying data-intensive flows. Specifically, in our study we aim at analyzing two main *scenarios* of data-intensive flows present in today's BI settings, namely:

- *extract-transform-load* (*ETL*). In the *traditional BI* setting, data are extracted from the sources, transformed and loaded to a target data store (i.e., a DW). For posing analytical queries (e.g., OLAP), the business users in such a scenario solely rely on the data transferred in periodically scheduled time intervals, when the source systems are idle (e.g., at night time).
- *extract-transform-operate* (*ETO*). *Next generation BI* has emerged as a necessity of companies for combining more instantaneous decision making with traditional, batched processes. In such a scenario, a user query, at runtime, gives rise to a data flow that accesses the data sources and alternatively crosses them with already loaded data to deliver an answer.

Kimball and Caserta [53] introduced the following definition of an ETL process "*A properly designed ETL system extracts data from the source systems, enforces data quality and consistency standards, conforms data so that separate sources can be used together, and finally delivers data in a presentation-ready format so that application developers can build applications and end users can make decisions.*"

Being general enough to cover the setting of data-intensive flows studied in this paper (i.e., both previous scenarios), we follow this definition and first divide our study setting into three main stages, namely:

(i) *Data extraction.* A data-intensive flow starts by individually accessing various (often heterogeneous) data sources, collecting and preparing data (by means of structuring) for further processing.
(ii) *Data transformation.* Next, the main stage of a data-intensive flow transforms the extracted data by means of cleaning it for achieving different consistency and quality standards, conforming and combining data that come from different sources.
(iii) *Data delivery.* This stage is responsible for ensuring that the data, extracted from the sources, transformed and integrated are being delivered to the end user in a format that meets her analytical needs.

Related fields. To complete the data-intensive flows lifecycle, in addition to the main three stages, we revisit two fields closely related to data-intensive flows, i.e., *data flow optimization* and *querying.*

(iv) *Data flow optimization* considers data-intensive flow holistically and studies the problem of modifying the given data flow, with the goal of satisfying certain non-functional requirements (e.g., performance, recoverability, reliability). Obviously, the optimization problem is critical for data flows in today's BI systems, where the data delivery is often required in the near real-time manner.

In addition, for the completeness of the overall picture, we briefly analyze what challenges the two main scenarios in data-intensive flows (i.e., *ETL* and *ETO*) bring to *querying*.

(v) The *requirements elicitation and analysis* (i.e., *querying*) stage mainly serves for posing analytical needs of end users over the available data. This stage is not actually part of a data-intensive flow execution, but depending on the scenario (i.e., either *ETO* or *ETL*), can respectively come as a preceding or subsequent stage for a data-intensive flow execution. At the lower level of abstraction, end users' analytical needs are typically expressed in terms of queries (e.g., SQL), programs (e.g., ETL/MapReduce jobs), or scripts (e.g., Pig Scripts), which are then automatically translated to data-intensive flows that retrieve the needed data. The typical challenges of *querying* the data in the *next generation BI* setting concern the ability of the system to adapt and complement users' analytical needs by means of discovering related, external data, and the usability of a BI system for posing analytical needs by end-users. The former challenge may span from the traditional DW systems that typically answer user's OLAP queries solely by exploiting the data previously loaded into a DW (by means of an ETL process), to situation-(context-)aware approaches that considering end user queries, explore, discover, acquire, and integrate external data [1,2]. Regarding the latter challenge, we can also observe two extreme cases: traditional querying by means of standard, typically declarative query languages (e.g., SQL, MDX), and approaches that enable users to express their (often incomplete) analytical needs in a more natural and human-preferred manner (e.g., keyword search, natural language). Recently, some researchers have proposed more flexible (semi-structured) query language (SQL++) for querying a variety of both relational and new NoSQL databases that may store data in a variety of formats, like JSON or XML [69].

Other approaches also tackled the problem of providing a higher level of abstraction for posing information requirements, more suitable for business users. As analyzing information requirements and systematically incorporating them into a design of data-intensive flows has been typically overlooked in practice, initial efforts were mostly toward systematic requirements analysis in BI and DW projects (e.g., [36,102]). However, such approaches still require long processes of collecting the requirements at different levels of abstraction and their analysis, before manually incorporating them into a data-intensive flow design. Thus, they obviously cannot be applied in *ETO* scenarios, where the generation of data-intensive flows to fulfill end user analytical needs is expected in near real-time. Other works identified such problem and proposed certain automation to such time-lasting process (e.g., [79]). However, the automation required lowering the level of abstraction for defining information requirements, tightening them to the multidimensional model notation (i.e., facts and dimensions of analysis). As an extension to this approach the authors further tried to raise the level and provide an abstraction of the data sources' semantics in terms of a domain ontology, with its graph representation. This has partially hidden the model specific details from business users, and allowed them to pose information requirements using a domain vocabulary. Recent work in [31] conducted an extensive study of decision support system approaches, identifying how they fit the traditional requirements engineering framework (i.e., [72]). As a result, the authors identified a need for systematic and structured requirements analysis process for further raising the level of automation for the design of decision support systems (including the design of data-intensive flows), while at the same time keeping the requirements analysis aware of all stakeholder needs. We have discussed here the main challenges that *requirements elicitation and analysis* introduces to the field of data-intensive flows in the *next generation BI* setting, but we omit further analysis as it falls out of the scope of this work.

As a result, we define a blueprint for studying data-intensive flows, depicted in Fig. 4. Going from top down, we depict separately the two main scenarios of data-intensive flows studied in this paper (i.e., ETL and ETO). Going from left to right, the first part of Fig. 4 (i.e., A, E) depicts the characteristics of the *data extraction* stage in terms of the complexity that different input data types bring to a data-intensive flow; then the following part (i.e., B, F) covers the characteristics of the *data transformation* stage; the penultimate part (i.e., C, G) covers the *data delivery* stage, while the last (most right) part (i.e., D, H) covers *querying*. Being rather a holistic field (i.e., taking into account the complete data-intensive flow), the *optimization* spans all stages of data-intensive flows and it is depicted at the bottom of Fig. 4.

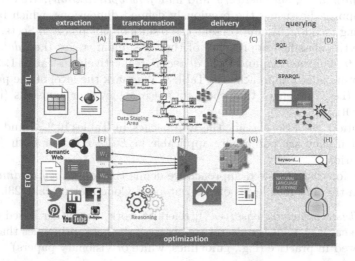

Fig. 4. Study setting for data-intensive flows (Color figure online)

3.3 Phase II (Analyzing the Characteristics of Data-Intensive Flows)

This phase included the review of the works that scrutinize the characteristics of data-intensive flows in both previously defined scenarios, i.e., *ETL* and *ETO*. This phase aimed at characterizing data-intensive flows in terms of the features that best indicate the movement toward the *next generation BI* setting.

To this end, we performed an incremental analysis of the included works to discover the features of data-intensive flows they have tackled. We started from the papers that individually covered the *traditional* and the *next generation BI* settings. The identified features are then translated into the dimensions for studying data-intensive flows (see Fig. 5). As new dimensions are discovered, the related papers are reconsidered to analyze their assumptions regarding the new dimensions. Each discovered dimension determines the levels, supported or envisioned by analyzed approaches, in which these approaches attain the corresponding feature of data-intensive flows. Eventually, we converged to a stable set of dimensions, which can be further used for studying and classifying the approaches of data-intensive flows.

In Sect. 4, we discuss in more detail the process of discovering dimensions for studying data-intensive flows, and further provide their definitions.

3.4 Phase III (Classification of the Reviewed Literature)

In this phase, we further extend the study to the works that more specifically cover the previously discussed areas of data-intensive flows (i.e., *data extraction*, *data transformation*, *data delivery*, and *data flow optimization*). We classify the reviewed approaches using the previously defined dimensions, which build our study setting (see Fig. 5), and present the results of this phase in Sects. 5–8. We summarize the classified approaches of the three main stages (i.e., *data extraction*, *data transformation*, and *data delivery*) respectively in Tables 1, 2 and 3, and the optimization approaches in Table 4. We mark the level of the particular dimension (i.e., challenge) that each approach achieves or envisions (i.e., Low, Medium, or High)[6].

In addition, we also classify the approaches in Tables 1, 2, 3 and 4 based on the fact if they are potentially applicable in *ETL*, *ETO*, or both *ETL* and *ETO* scenarios.

Finally, for each reviewed approach we define the *technology readiness level*, focusing on the first four levels of the European Commission scale [23].

- TRL1 (*"basic principles observed"*), refers to work that either based on practical use cases or reviewed literature observes the basic principles that should be followed in practice (e.g., guidelines, white or visionary papers)

---

[6] The exception to this are the approaches from the data flow optimization area, for which we introduced levels that more precisely describe the consequences of their placement inside the corresponding dimensions. Moreover, in the cases when the approach is completely independent of the level for a particular dimension, we mark it as non-applicable (N/A).

- TRL2 (*"technology concept formulated"*), refers to work that provide theoretical underpinnings of the studied area, which are not always directly applicable in practice, but represent an important foundation for principles that should be followed in practice (e.g., the database theory works on data exchange and data integration). ·
- TRL3/TRL4 (*"experimental proof of concept"*/ *"technology validated in lab"*), refers to the system-oriented work that provide the proof of concept solution for an observed principle from the previous two levels, validated either over synthetic (TRL3) or real-world use cases (TRL4).

4 Defining Dimensions for Studying Data-Intensive Flows

For each area in the outlined study setting for data-intensive flows (Fig. 4), we discuss in more detail, and further provide the definitions of the dimensions through which the reviewed works on data-intensive flows are analyzed (see Fig. 4). Then, in the following Sects. 5–8, we discuss in more detail the works specifically covering each of the studied areas.

4.1 Data Extraction

The most commonly discussed challenge when extracting data (e.g., [1,53]) is related to the format in which the data are provided, i.e., *structuredness*.

Structuredness determines the level, in which data in data sources under analysis follow a certain model, constraints or format. It spans from highly structured data that follow strictly defined models and ensures certain constraints over data (High), like relational (see top left of Fig. 4); then semi-structured data that are represented in a repetitive [47], standard and easily parsable format, but that do not enforce strong constraints over data (Medium), like XML,

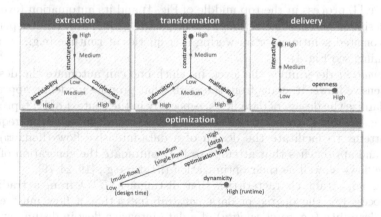

Fig. 5. Dimensions for studying data-intensive flows

CSV, RDF (see middle left in Fig. 4); and unstructured data in a free-form (textual or non-textual) that require smarter techniques for extracting real value from it (`Low`), like free text, photos, x-rays, etc. (see bottom left of Fig. 4).   □

Other characteristics of this stage are related to the degree in which BI applications need to be coupled with source systems when extracting the data, i.e., *coupledness*, and the reliability of accessing these systems, i.e., *accessibility*.

Coupledness determines the level, in which data-intensive flows depend on a specific knowledge or components obtained from data sources under analysis. It spans from typical ETL processes that consolidate organization's ("in-house") operational sources with predictable access policies (e.g., DBMS, ERP systems; see top left of Fig. 4), using extraction algorithms (e.g., incremental DB snapshots) that strongly depend on information from source systems (`High`), e.g., materialized views DW solutions, triggers, source components modifications; then the systems that use logs, timestamps or other metadata attached to data sources (`Medium`); and the scenarios where we cannot expect any in-advance knowledge about the data sources, but the system needs to discover and integrate them on-the-fly [2] (`Low`), e.g., Web data, Link Open Data.   □

Accessability determines the level, in which one can guarantee a "non-stop" access to certain data sources. It spans from "in-house", highly available data (`High`), like ERP, ODS systems; then the external data usually provided by data providers that under SLAs can guarantee certain accessibility to data (`Medium`); and external data sources that are completely out of the organization's control (`Low`), like open, situational, or Web data.   □

4.2   Data Transformation

In this stage, the most common issue is related to the complexity of data transformations inside a flow (e.g., [96]), and more specifically to the degree in which we can automate the design of a data-intensive flow, i.e., *automation*. While the design of an ETL process is known to be a demanding task (e.g., [89]; see the example ETL process in the top middle of Fig. 4) and its automation (even partial) is desirable and has been studied in the past (e.g., [68,79]), ETO depends on fully automated solutions for answering user queries at runtime (e.g., by means of reasoning; see Fig. 4).

Automation determines the level, in which one can automate the design of data-intensive flows. It spans from the works that propose modeling approaches to standardize the design of data flows, especially in the context of ETL processes (`Low`), e.g., [3,92,99]; then approaches that provide guidelines and/or frequently used patterns to facilitate the design of a data-intensive flow (`Medium`), e.g., [53,98]; and approaches that attempt to fully automate the generation of data-intensive flows as well as their optimization (`High`), e.g., [19,24,79].   □

Other important characteristics that distinguish *ETO* from a traditional *ETL* process are the degree of data *constraintness* that a flow must ensure, and the flexibility (i.e., *malleability*) of a data-intensive flow in dealing with the changes in the business context.

Constraintness determines the level, in which data-intensive flows must guarantee certain restrictions, constraints, or certain level of data quality over the output data. It spans from fully constrained data, usually enforcing the MD model (MD integrity constraints [64]), and high level of data cleanness and completeness required to perform further data analysis, like OLAP (High); then, data-intensive flows that may provide ad-hoc structures for answering user queries (e.g., reports), without a need to enforce the full completeness and cleanness of data (Medium); and as an extreme case we consider the concept of *data lakes* where no specific schema is specified at load time, but rather flexible to support different analysis over stored and shared data at read-time (Low).   □

Malleability determines the level in which a system is flexible in dealing with the changes in the business context (e.g., new/changed data sources under analysis, new/changed/removed information requirements). It spans from the traditional DW settings where data sources as well as information requirements are static, typically gathered in advance. and are added manually to the analysis only at design time, while any change would require the redesign of a complete data-intensive flow (Low) [53]; then systems that tackle the incremental evolution of data-intensive flows in front of new requirements and data sources (Medium) [49]; and dynamic approaches that consider discovering new, usually external data at runtime (High) [2].   □

4.3 Data Delivery

For delivering the transformed data at the target, we have identified two main characteristics that distinguish ETL from ETO, namely the *interactivity* of data delivery as perceived by the end-user; and the *openness* of the delivered data, which refers to the degree in which the approaches assume the delivered information to be complete (i.e., *closed* vs. *open world assumption*).

Interactivity determines the level in which a system interacts with the end-user when executing a data-intensive flow and delivering the data at the output. It spans from traditional ETL processes that typically deliver the data (i.e., materialize the complete data to load a DW) in a batched, asynchronous process, without having an interaction with an end-user (Low), then approaches that based on the overall cost, select data to be partially materialized (e.g., loaded in a batch to materialized views), and those that are queried on-the-fly (Medium); and finally completely interactive approaches that assume on-the-fly data flows which deliver the data to answer user queries for immediate use only, e.g., for visualization (High).   □

Openness determines the level, in which the delivered data are considered *open* to different interpretations. It spans from *closed-world assumption* approaches typical for traditional databases, where the data are considered complete and any answer to a user query is determined (Low), to *open-world assumption* approaches, where due to the assumption that data may be incomplete, an answer to a user query can be either determined if there exist data that can prove such an answer, or "unknown" in the case where there is no data to determine its truthfulness (High).   □

4.4   Optimization of Data-Intensive Flows

Finally, we also discuss the low *data latency* as an important requirement for today's data-intensive flows.

Optimizing data flows has been one of the main topics in database research [48]. In this context, we discuss the following two aspects.

Optimization input. This dimension refers to the level at which the optimization is provided. It spans from optimizing the way that input `data` is stored and processed (e.g., data fragmentation) in order to achieve optimal execution (e.g., parallelizing data flow execution); then optimizing the execution of `single data flows` by means of modifying the execution of the data flow (e.g., operation reordering, different implementations) to achieve the optimal execution of a data flow; and finally the overall optimization of a `multi-flow`, where the goal is to achieve the optimal execution for a set of data flows, rather than optimizing the execution of a single flow.   □

Dynamicity. Another dimension for studying the optimization of data-intensive flows relates to the overhead introduced by the optimization and thus determines the level of *dynamicity* of the data flow optimization process. In the traditional DW systems, the design and optimization of ETL processes is done at the `design time`, once while the process is then executed periodically, many times. This obviously allows for a higher overhead of the optimization process and taking into account different metadata (e.g., statistics from the previous executions of the same flow). On the other hand, an *ETO* flow must be optimized at `runtime`, when the analytical query is issued, which introduces additional challenges into the optimization techniques, especially regarding the way the flow and data statistics are gathered and exploited for optimization.   □

5   Data Extraction

In the initial stage of data-intensive flows, the source systems are identified and accessed for extracting the relevant data. In the data extraction stage, we focused on analyzing the three following challenges that characterize the shift towards the *next generation BI* settings, namely *coupledness*, *accessibility*, and *structuredness*, which are subsequently discussed in the following subsections.

5.1   Structuredness

The seminal work on DW systems (e.g., [46]), although mentioned external unstructured data as an important opportunity for building DW system, have not in particular tackled the challenges that they bring to the design of the ETL pipelines. Furthermore, some of the first (purely relational) approaches for the DW system design in the literature (i.e., using materialized views in [90]), as well as the later extraction techniques (e.g., [57,62]) assumed purely relational data sources, thus only supported `High` structuredness of input data. [53] considers

`Medium` stucturedness of input data, by providing the practical guidelines for accessing external data in Web logs or flat files, and semi-structured data (i.e., XML; see Fig. 4(A)).

Example. In our running example, the ETL flow depicted in Fig. 4 reads data from the transactional systems supporting daily sales operations. Notice that besides the challenges that heterogeneity in data sources (both structural and semantic [87]) brings to the design of an ETL pipeline, well-structured data sources do not require any special technique for obtaining the data before the transformation and cleaning starts. However, today, using a diversity of external and often unstructured data sources is becoming inevitable and thus the techniques for extracting such data have attracted the attention of both academy and industry. In our example scenario, we can see that introducing unstructured data (e.g., free text `reviews` from the Web) into the analytical processes of the company (see Figs. 2 and 3) required additional data processing for extracting relevant information from these sources. Specifically, `Sentiment Analysis` based on natural language processing (NLP) is performed over textual data from the Web to extract customer opinions about the items and the campaign.   □

In research, different techniques (e.g., text mining [28], NLP [29,58], sentiment analysis) are proposed to discover data patterns and extraction rules for extracting relevant data from natural language documents and transform unstructured data into more explicitly structured formats [10] (e.g., graphs, trees, relational model). There, approaches are hence able to deal with the `Low` structuredness of input data. However, at the same time, they may assume a considerable latency overhead to the execution of the complete data pipeline and thus introduce additional challenge to data-intensive flows. Interestingly, the linked data movement [7], on the other side, proposes that large amounts of external data are already provided in more structured (semi-structured) formats and semantically interlinked (e.g., using RDF), in order to facilitate the situation-aware analysis [63] and data exploratory actions [2]. These approaches assume `Medium` structuredness of input data, ready for the analytical processes carried out by data-intensive flows.

5.2   Coupledness

First relational approaches for designing a DW by means of a set of materialized views (e.g., [90]) in general allowed very efficient refreshments processes, by applying the well-known view maintenance techniques, to either compute incremental changes in the sources or a complete "rematerialization" of a view. Both approaches issue queries (*maintenance queries*) over the data sources, extract the answer, and load the corresponding views. Such approaches, however, required a `High` coupledness to source systems and soon became unrealistic to support the demands of enterprises to include a variety of external data sources into their analytical processes.

Example. As we can see from our example scenario, the retail company initially relied mainly on the internal data sources (e.g., the information about the `sold` items), which are periodically transferred to a central DW (see Example 1.1). To lower the data transferred in every execution of the ETL flow, the designers have built the flow to only extract the `sales` and `item` data that are inserted to the sources after the last ETL execution (i.e., snapshot difference). For efficiently finding the differences in two snapshots of the source data, the tight (`High`) coupledness to the considered data sources is needed. On the other side, in the scenario illustrated in Examples 2.1 and 2.2 (i.e., Figs. 2 and 3, respectively), some data sources (i.e., `item reviews` from the Web) are not under the control of the company and moreover they may not be known in advance as their choice depends on the current user needs. Thus, obviously we cannot depend on having strong knowledge of these kinds of data sources.   □

In the context of modern ETL processes, in [96], the author revisits the approaches for finding the difference of the consecutive snapshots of source databases (e.g., [57,62]). However, these snapshot techniques (e.g., timestamps, triggers, interpreting the source's logs) still required certain control over the known source systems (i.e., `Medium` coupledness). Web-oriented systems have further imposed more relaxed environments for extracting data located on disparate Web locations. The most common solution introduced for performing data integration from disparate sources includes building specialized software components, called *wrappers*, for extracting data from different Web sources (e.g., [32]; see Fig. 4(E)). Wrappers are typically used in combination with another component (namely *mediator*, see Fig. 4(F)), which, based on a user query, invokes individual wrappers, and combines (i.e., integrates) data they return to answer the input query. However, the wrapper/mediator architecture still requires a `Medium` coupledness, as wrapper design highly relies on the specific technology of data source systems, while the changes in the source systems typically require reconsidering the wrappers' design. Finally, as we mentioned above, to make the enormously growing data volumes on the Web available and potentially useful for the enterprise analytical and data discovery actions, the linked and open data movement (e.g., [7]) has proposed a completely uncoupled environment (i.e., `Low` coupledness) with the general idea of having huge amounts of distributed data on the Web semantically interlinked and preferably provided in easily parseable formats (e.g., XML, RDF), see Fig. 4(E). Approaches that argue for such `Low` coupled scenarios, envision architectures that can take the advantage of existing logic-based solutions for enabling data exploration actions over the external sources [2], and provide more context-aware data analysis [1].

A separate branch of realated research that strongly argues for `High` decoupling of data sources in data-intensive flows is Complex Even Processing (CEP) [14]. The idea here is on enabling on-the-fly processing and combining of data coming in greater speed and typically from external data sources, with the goal of detecting different correlations or anomalies happening in the "external world". Thus, the CEP systems typically decouples from the technical level information of the data sources (e.g., sensor readings), and rather aims at detecting events at the application level (e.g., correlations, anomalies). CEP is rather extensive and separate fields of research, and to this end, we here give its high level overview in terms of our analysis dimensions, while for the specific approaches and applications we refer the reader to the survey in [14], which compares and studies in detail the state of the art approaches in this field.

5.3   Accessability

The practical guidelines for efficiently building an ETL process in [53] proposes a pre-step of profiling data sources for quality, completeness, fitness, and accessibility. Apart from transactional data sources, dominant in traditional DW scenarios [46], with typically `High` accessability or at least predictable access policies (e.g., nightly time windows), nowadays a vast amount of potentially useful data for an enterprise is coming from remote data sources, over the global networks, like forums, social networks, and Web in general, [1], see Fig. 4(E).

Example. Going back to our example scenario, we can notice that in the traditional DW environment, the company builds the system based on the previously elicited business needs and accordingly incorporates internal data sources (e.g., `item` and `sales`) into their analytical processes. ETL flows (e.g., see Fig. 1) are designed and tested in advance for periodically extracting the data from pre-selected data sources, relying on the `High` or predictable availability of these sources. Conversely, the ETO flows in Figs. 2 and 3 cannot rely on accessing the data sources at all times, due to remote access (e.g., Web and social networks) and moreover as they can be selected on-the-fly.   □

Even though in the linked (open) data movement information about quality of external data are envisioned to be in the form of catalogs [7], the access to these data at any moment still cannot be guaranteed (`Low` accessability), which brings a new challenge to the process of data extraction in this scenario. In this context, the authors in [1] study the concept of situational data, which are usually external to an organization control and hence without a guaranteed access, but which in fact play an important role in today's context-aware decision making. The authors thus propose a *"data as a service"* solution, where envisioned systems will have a registry of possible data providers, and using Web service interface partially automate the process of finding the most appropriate and currently accessible data source.

5.4   Discussion

We summarize the results of studying the challenges of *data extraction* stage (i.e., the classification of the representative approaches) in Table 1. As expected, we have found more matured (i.e., $TRL \geqslant 2$) works dealing with this stage in the *traditional BI* setting (i.e., *ETL*), considering tighter *coupledness* to source systems, relying on high *accessibility*, and expecting *structured* data. Several approaches have opened the issue of dynamically accessing external and unstructured data, focusing mostly on data coming from the Web, while the majority considered structured (relational) or at most semi-structured (XML) data.

Table 1. Classification of data extraction approaches

Data extraction		TRL	Dimensions		
ETL vs. *ETO*	Approaches				
	AUTHORS, YEAR, [*NAME*,] REFERENCE		*struct.*	*access.*	*coupl.*
ETL	Inmon, 1992, [46]	1	High	High	High
	Theodoratos & Sellis, 1999, [90]	2			
	Labio et al. 1996, [57]	3	High	N/A	Medium
	Lindsay et al. 1987, [62]				
	Kimball & Caserta, 2004, [53]	1	Medium	Medium	High
ETO	Feldman & Sanger, 2007, [28]	1	Low	N/A	N/A
	Buneman et al., 1997, [10]				
	Laender et al., 2002, [58]	1	Low	Low	Medium
	Bizer et al., 2009, *Linked Data*, [7]	1	Medium	Low	Low
	Cugola and Margara, 2012, *CEP*, [14]				
	Abelló et al., 2013, *Fusion Cubes*, [1]	1	Low	Low	Low
	Abelló et al., 2015, [2]				
	Garcia-Molina et al., 1997, *TSIMMIS*, [32]	4	Medium	High	Medium

Data extraction is however an important stage to be reconsidered for today's data-intensive flows, especially taking into account new loosely coupled BI environments [1]. The main challenges of these (mostly envisioned) ecosystems with low coupledness relate to the fact that data sources are outside of the organization control, and often not even known in advance. Thus, the efficient techniques to discover the relevant data must be deployed. We can benefit here from the known techniques proposed to explore the contents on the Web (e.g., *Web crawling*). Moreover, being external data sources, the systems become very sensitive to very probable variability of data formats, as well as the undefined semantics of data coming from these sources. To overcome the semantics heterogeneity gap between the data sources, and to automate discovering and extracting the data from them, we propose to use the semantic-aware exploratory mechanisms [2].

Furthermore, as we can see from our example scenario, specialized data processing (e.g., *natural language processing, sentiment analysis, text mining*) should be also considered to extract the relevant information from these, often unstructured, data sources. However, such complex data transformations typically affect the data latency in a data-intensive flow, hence in most of the current approaches this kind of input data transformation has been considered as part of a pre-processing step. An alternative to this, following the principles of *linked (open) data* [7], is to have data published in at least semi-structured formats (e.g., XML, CSV, RDF), which largely facilitates their further exploitation.

6   Data Transformation

After data are extracted from selected (often heterogeneous) sources, the flow continues with transforming the data for satisfying business requirements and considered quality standards. Data transformation is characterized as the main stage of a data-intensive flow by most of the approaches [53, 96]. The main dimensions we analyze in the data transformation stage are *automation, malleability*, and *constraintness*.

As previously mentioned, from early years, managing heterogeneous data has brought the attention of database community, and some fundamental works stem from the database theory field (i.e., *data integration* (e.g., [59, 94]) and *data exchange* (e.g., [27])) to tackle this problem from different perspectives.

Both data exchange and data integration problems are based on the concept of *schema mappings*, which in general can be seen as assertions that define the correspondences between source and target schema elements.

In general, a parallelism can be drawn between the theoretical problems of data exchange and data integration, and what we today consider as data-intensive flows. Similar observation has been discussed in [96]. The author compares data exchange to the traditional DW setting, where data transformations in the ETL pipeline can be generally seen as schema mappings of the data exchange setting. However, as also noted in [96], schema mappings, as defined by these theoretical approaches, are typically limited to simple transformations over data and do not efficiently support typical transformations of data-intensive flows (e.g., grouping, aggregation, or black-box operations), nor the diversity of data sources (i.e., only relational or XML data formats have been considered).

6.1   Malleability

Data-intensive flows, as other software artifacts, do not lend themselves nicely to evolution events, and in general, maintaining them manually is hard. The situation is even more critical in the *next generation BI* settings, where on-the-fly decision making requires faster and more efficient adapting to changed domain context, i.e., changed data sources or changed information needs.

For considering the former problem, we revisit the foundational works on data exchange and data integration, which introduced two main approaches for schema mappings, i.e., *global-as-view* (GAV) and *local-as-view* (LAV) [30, 59].

In the GAV approach, the elements of the global schema are characterized in terms of a query over the source schemata, which further enables less complex query answering by simply unfolding global queries in terms of the mapped data sources. However, GAV mappings lack flexibility in supporting the evolution of data source schemata, as any change on the sources may potentially invalidate all the mapping assertions (i.e., Low malleability). An example of this approach is the wrapper/mediator system [32].

Example. As we discussed, in Scenario 1, the company elicits the business needs prior to designing the DW and ETL flows (e.g., see Fig. 1). In the case a new data source is added, the redesign of the system is performed offline before the ETL flows are run again. However, notice that in the second scenario (see Examples 2.1 and 2.2) the flexibility of the system for adding new data sources must be supported in an "instant" manner, as business needs are provided on-the-fly and often require a prompt response.   □

As opposed to GAV, LAV schema mappings characterize the elements of source schemata in terms of a query over the global schema. LAV mappings are intuitively used in the approaches where changes in dynamic data source schema are more common (e.g., [54]) as it provides High malleability of the data integration systems. We can thus observe that the LAV approach fits better the needs of the *next generation BI* setting, where the variability and number of data sources cannot be anticipated (e.g., [1,2]). However, the higher flexibility of LAV mappings brings the issues of both, the complexity of answering the user queries and the completeness of the schema mappings. Generally, in LAV, answering user queries posed in terms of a global schema implies the same logic as answering queries using materialized views, which is largely discussed as a computationally complex task [42].

Several approaches further worked on generalizing the concept of schema mappings by supporting the expressive power of both LAV and GAV, i.e., *both-as-view* (BAV) [65], and *global-and-local-as-view* (GLAV) [30].

However, as we discussed before such approaches are hardly applicable to the complex data-intensive flows. In the context of the traditional DW systems, some works have studied the management of data-intensive flows (i.e., ETL process) in front of the changes of data source schemata. In [70] the authors propose a framework for impact prediction of schema changes for ETL workflow evolution. Upon the occurred change, the ETL flow is annotated with (pre-defined) actions that should be taken, and the user is notified in the case that the specific actions require user involvement. Other approaches (e.g., [49]) have dealt with automatically adapting ETL flows to the changes of user's information needs. For each new information requirement, the system searches for the way to adapt the existing design to additionally answer the new requirement, by finding the maximal overlapping in both data and transformation. Lastly, some approaches have also dealt with the problem of adapting DW systems to the changes of the target DW schema. Being a "non-volatile collection of data" [46], the evolution changes of the target DW schema are typically represented in terms of different versions of a DW (i.e., multiversion DW). In particular, the most important issue was providing a transparent querying mechanisms over different versions of DW schemata (i.e., cross-version querying; [37,67]). For instance, a solution proposed in [37] suggested keeping track of change actions to further enable answering the queries spanning the validity of different DW versions. These approaches provide a certain (Medium) level of malleability for data-intensive flows, but still lack the full automation of the evolution changes or applicability in the case of unpredictable complexity of data transformations.

In addition, after data warehousing was established as a de facto way to analyze historical data, the need for more timely data analysis has also emerged in order to support prompter detection of different anomalies coming from data. This led researches to rethink the current DW architecture and make it more malleable to combine both traditionally mid-term and long-term, with "just-in-time" analysis. This brought the idea of (near) real-time data warehousing systems. Several approaches discussed the main requirements of such systems and proposed architectural changes to traditional DW systems for satisfying these new requirements (e.g., [8,97]). For instance, besides the main requirement of data freshness, [97] has also indicated minimal overhead of the source system and scalability in terms of input data sources, user queries, and data volumes, as relevant for these systems. They however pointed out the contradiction between users need for maximal data freshness and completeness, and the high overhead of the traditional DW workflows that often require costly data cleaning and transformations. To this end, the approaches in [8] and [97] discuss both conceptual and technological changes that would balance the delays in traditional ETL processes. In practice, SAP Business Warehouse [66] is an example of such system. It provides certain flexibility to traditional DW systems for enabling on-the-fly analysis at different levels of data, i.e., summarized and loaded to a DW, consolidated operational data (operational data store), or even directly over the transactional data. Their goal is to enable more real-time data warehousing and a possibility of also including fresh, up-to-date transactional data to the analysis. Even though the (near) real-time DW approaches bring more malleability (Medium) to data analysis by combining historical and on-the-fly analysis, included data are still typically coming from the in-house and predefined data sources.

6.2   Constraintness

What further distinguishes data exchange [27,55] from the original data integration setting, is that the target schema additionally entails a set of constraints that must be satisfied (together with schema mappings) when creating a target instance (i.e., High constraintness).

Example. In Fig. 1, we can notice that for loading data into a DW, data-intensive flow must ensure a certain level of data quality to satisfy constraints entailed by the DW (e.g., Remove Duplicates in Fig. 1 removes the repetitive itemIDs for loading the successFact table into a DW). On the other side, data-intensive flows in the *next generation BI* settings (see Figs. 2 and 3), due to their time constraints typically cannot afford to ensure full data quality standards, but is often sufficient to deliver partially cleaned (i.e., "right") data, at the right time to an end-user [22].   □

The work on generalizing schema mappings (GLAV) in [30] also discusses the importance of adding support for defining the constraints on global schema, but no concrete solution has been provided. In the data integration setting, although some works did study query answering in the presence of integrity constraints on global schema [12], (i.e., Medium constraintness), most of the prominent data integration systems (e.g., [32,54]) typically do not assume any constraints in the target schema (i.e., Low constraintness). Furthermore, as we discussed in the DW context, the design of an ETL process is affected by the integrity constraints typical in a dimensionally modeled DW schema (see Fig. 4(C)).

When working with data from unstructured data sources, one may face two different problems: (1) how to extract useful information from data in unstructured formats and create more structured representation; and (2) how to deal with incomplete and erroneous data occurred due to lack of strict constraints in source data models. The latter problem becomes even more challenging when the target data stores entail strict constraints as we discussed above. While the former problem is usually handled by means of data extraction techniques discussed in the Sect. 5, the latter is solved at the data transformation stage, where data are cleaned to fulfill different quality standards and target constraints. As expected, such a problem has brought the attention of researchers in the data exchange (e.g., [25]) and data integration (e.g., [21]) fields. In the modern DW systems, target data quality and `High` constraintness is usually guaranteed as the result of the process called data cleaning. Data cleaning deals with different data quality problems detected in sources, e.g., lack of integrity constraints at sources, naming and structural conflicts, duplicates [75].

6.3   Automation

It is not hard to see from the previously discussed problem of data cleaning and the flows in the example scenario, that today's data-intensive flows require more sophisticated data transformations than the ones (mainly based on logics) assumed by fundamental approaches of *data exchange* and *data integration*. At the same time, higher automation of the data flow design is also required to provide interactive, on-the-fly, analysis.

Example. Loading a DW may require complex data cleaning operations to ensure the entailed constraints. Obviously, complete automation of the design of such data-intensive flows is not realistic and thus the designers in Scenario 1 usually rely on a set of frequent data transformations when building ETL flows (e.g., `Join`, `UDF`, and `Remove Duplicates` in Fig. 1). But, in Scenario 2, such an assisted design process is not sufficient, as the flows for answering users' on-the-fly queries (e.g., see Examples 2.1 and 2.2) must be created instantaneously. This, together with the requirement for lower data latency, restricted such flows to more lightweight operations (e.g., `Filter` or `Match` in Fig. 2).   □

Different design tools are available in the market and provide often overlapping functionalities for the design and execution of data-intensive flows (mostly ETL; see for example Fig. 4(B)). The complexity and variability of data transformations has introduced an additional challenge to the efforts for providing a commonly accepted modeling notation for these data flows. Several works have proposed different ETL modeling approaches, either ad-hoc [99], or based on well-known modeling languages, e.g., UML in [92] or BPMN in [3,101]. However, these modeling approaches do not provide any automatable means for the design of an ETL process (i.e., `Low` automation). Some approaches (e.g., from UML [68], or from BPMN [4]) are further extended to support certain (i.e., `Medium`) automation of generating an executable code from the conceptual flow design, by means of model transformations (i.e., Model-driven design).

The design of an ETL process is on the other side described as the most demanding part of a DW project. As reported in [89] ETL design can take up to

80 % of time of the entire DW project. In [53] the authors give some practical guidelines for a successful design and deployment of an ETL process, but without any automatable means (i.e., Low automation), still, a considerable manual effort is expected from a DW designer. In [98], the framework that uses the ad-hoc modeling notation from [99] is proposed to assist the design of ETL processes, along with the palette of frequently used ETL patterns (i.e., Medium).

Several approaches went further with automating the conceptual design of ETL processes. On the one hand, in [88], the authors introduced the design approach based on Semantic Web technologies to represent the DW domain (i.e., source and target data stores), showing that this would further enable automation of the design process by benefiting from the automatic reasoning capabilities of an ontology. [79], on the other hand, assumes that only data sources are captured by means of a domain ontology with associated source mappings. Both DW and ETL conceptual designs are then generated to satisfy information requirements posed in the domain vocabulary (i.e., ontology). Finally, [5] entirely rely on an ontology, both for describing source and target data stores, and corresponding mappings among them. Integration processes (ETL) are then also derived at the ontological level based on the type of mappings between source and target concepts (e.g., equality, containment). However, even though these approaches enable High automation of the data flow design, they work on a limited set of frequent ETL operations.

In parallel, in the field of data exchange, [24] proposes a tool (a.k.a. *Clio*) that automatically generates correspondences (i.e., schema mappings) among schemas without making any initial assumptions about the relationships between them, nor how these schemas were created. Such a generic approach thus supports High automation in creating different schema mappings for both data integration and data exchange settings. [19] went further to provide the interoperability between tools for creating declarative schema mappings (e.g., *Clio*) and procedural data-intensive tools (e.g., ETL). Still, such schema mappings either cannot tackle grouping and aggregation or overlook complex transformations typical in today's ETL processes. The *next generation BI* settings, however, cannot always rely on the manual or partially automated data flow design. Moreover, unlike ETL, ETO cannot completely anticipate end user needs in advance and thus besides the High level of automation, the design process must also be agile to efficiently react in front of new or changed business needs (e.g., [78]).

6.4   Discussion

As we have seen, in the *next generation BI* setting (i.e., *ETO*), where data sources are often external to the organization control and moreover discovered dynamically based on current user needs; a more flexible environment is needed for efficiently supporting adding new data sources to analytical processes. The local-as-view (LAV) schema mapping approaches are more suitable in order to support the required level of *malleability* [30,65]. In the LAV approach, plugging new data sources requires defining a mapping of the new source schemata to the global schema, without affecting the existing mappings. However, as we can see

Table 2. Classification of data transformation approaches

Data transformation		TRL	Dimensions		
ETL vs. ETO	Approaches				
	AUTHORS, YEAR, [NAME,] REFERENCE		autom.	malleab.	constr.
ETL	Fagin et al., 2003, *Data Exchange*, [27]	2	N/A	N/A	High
	Kolaitis, 2005, [55]				
	Rahm & Hai Do, 2000, [75]	1			
	Kimball & Caserta, 2004, [53]	1	Low	Low	High
	Vassiliadis et al., 2002, [99]	2	Low	Low	High
	Trujillo & Luján-Mora, 2003, *UML-ETL* [92]				
	Wilkinson et al., 2010, *xLM*, [101]				
	El Akkaoui et al., 2012, *BPMN-ETL* [3]				
	Muñoz et al., 2009, [68]	3	Medium	Medium	High
	El Akkaoui et al., 2013, [4]				
	Vassiliadis et al., 2003, *ARKTOS II*, [98]				
	Papastefanatos et al., 2009, [70]				
	Morzy & Wrembel, 2004, [67]				
	Golfarelli et al., 2006, [37]				
	Skoutas & Simitsis, 2007, [88]	3	High	Low	High
	Bellatreche et al., 2013, [5]				
	Fagin et al., 2009, *Clio*, [24]	4			
ETL & ETO	McDonald et al., 2002, *SAP BW*, [66]	4	Medium	Medium	High
	Romero et al., 2011, *GEM*, [79]	3	High	Medium	High
	Dessloch et al., 2008, *Orchid*, [19]				
	Jovanovic et al., 2012, *CoAl*, [49]				
ETO	Garcia-Molina et al., 1997, *TSIMMIS*, [32]	4	High	Medium	Low
	Kirk et al., 1995, *Information Manifold*, [54]	3	High	High	Low
	Romero & Abelló, 2014, [78]	1			
	Abelló et al., 2014, [2]				
	McBrien & Poulovassilis, 2003, *BAV*, [65]	2	N/A	High	Medium
	Friedman et al., 1999, *GLAV*, [30]				
	Calì et al., 2004, [11]				

in Table 2, currently, most of these techniques are still purely theoretical (i.e., $TRL = 2$), while the high complexity and intractability of LAV approaches have been widely discussed, and hence represent a serious drawback for using LAV mappings in near real-time BI environments.

On the other side, some approaches (e.g., [24]) have worked on automating the creation of such schema mappings, which can be widely applicable for supporting answering information requirements on-the-fly. Even though we notice the lower requirement for the cleanness and constraintness of output data in the *next generation BI* setting (see Table 2), which would typically result with lower complexity of a data flow, today's BI applications do require rather complex data analytics, which are typically not supported in the schema mapping approaches. Some approaches try to extend this by automating the flow design (e.g., [5,79]), but still with very limited and predefined operation sets (i.e., *ETL & ETO*). Therefore, automating the creation of more complex data-intensive flows (e.g., machine learning algorithms), by means of exploiting different input data characteristics or using metadata mechanisms is still lacking.

7 Data Delivery

After data from multiple sources are cleaned, conformed, and combined together, a data-intensive flow delivers the data in the format suitable to the user needs either for visualization, or further analysis and querying. In the data delivery stage, we focus on analyzing the two following dimensions, namely, *interactivity* and *openness*, subsequently discussed in the following sections.

7.1 Interactivity

One of the important decisions that should be made while building data-intensive flows is the interactivity of the system when delivering data at the output.

Example. Notice that the ETL flow in Fig. 1 is designed to periodically transfer the complete data about the item sales in a batched back-end process, so that users may later analyze the subset of these data depending on their needs (e.g., slicing it only to the sales in the third quarter of the last). ETO flows in Figs. 2 and 3, however, instantly deliver the data from the sources that are currently asked by the user (e.g., trends of the past weekend, and trending product items from the first week of March, respectively). Moreover, such ETO flows are typical examples of answering ad-hoc and one-time analytical queries, thus storing their results is usually not considered as beneficial.   □

Going back again to the fundamental work on *data exchange*, the data delivery in this setting is based on computing a solution, i.e., a complete instance of the target schema that satisfies both, schema mappings and constraints of the target schema. The queries are then evaluated over this solution to deliver the answer to the user. However, due to incompleteness of data and/or schema mappings, there may be more than one, and theoretically an infinite number of valid solutions to the data exchange problem [27]. Answering user queries in such a case would result in evaluating a query over all possible solutions (i.e., finding the *certain answer*). To overcome the obvious intractability of query answering in data exchange, a special class of solutions (*universal solutions*), having a homomorphism into any other possible solution, is proposed.

Like in the data exchange setting, materializing the complete data at the target for the purpose of later answering user queries without accessing the original data sources, is also considered in the later approaches for designing a data warehouse (see Fig. 4(G)), i.e., Low *interactivity*. As we discussed in Sect. 5, a DW has been initially viewed as a set of materialized views (e.g., [90]). Similarly, in this case the database specific techniques (e.g., incremental view maintenance) are studied to minimize the size of the data materialized in each run of refreshment flows (maintenance queries). However, as DW environments have become more demanding both considering the heterogeneity and volume of data, it became unrealistic to consider a DW solely as a set of materialized views. In addition, many approaches have further studied the modeling and the design of a target DW schema, which should be able to support analytical needs of end users. This has resulted in the field of multidimensional (MD) modeling that is based on fact/dimension dichotomy. These works belong to a broader field of MD modeling and design that is orthogonal to the scope of this paper, and thus we refer readers to the survey of MD modeling in [77] and the overview of the current design approaches covered by Chap. 6 in [38].

Conversely, the data integration setting, as discussed in Sect. 6, does not assume materializing a complete instance of data at the target, but rather interactively answering individual user queries (e.g., through a *data cube* or a *report*; see Fig. 4(G)) posed in terms of a global (virtual) schema (i.e., High interactivity). A query is reformulated at runtime into queries over source schemata, using schema mappings (e.g., [32,42,54,59,94]). Another examples of High interactivity are Complex Event Processing and Data Stream Processing systems. Besides the differences these systems have (see their comparison in [14]), the common property of these systems is that they provide on-the-fly delivery of data, with typically low latency, and for the one-time use only (e.g., monitoring stocks, fraud detection), without a need to materialize such data.

In [34], in the context of peer data management, a hybrid solution is proposed based on both data exchange and data integration. The Medium *interactivity*, by partially materializing data and using a virtual view over the sources (peers), has been proposed. To this end, the solution presents schema dependencies that can be used both for computing the core and query answering.

The "right" (Medium) level of *interactivity* is also discussed to be crucial in the *next generation BI* setting (e.g., [15]) where a partial materialization is envisioned for a subset of data with low latency and low freshness requirements (i.e., for which we can rely on the last batched ETL run). Following the similar idea, [74] proposes a framework for combining data-intensive flows with user query pipelines and hence choosing the optimal materialziation point (i.e., Medium interactivity) in the data flow, based on different cost metrics (e.g., source update rates and view maintenance costs). Another field of research also follows the Medium level of *interactivity*, and proposes an alternative to traditional ETL processes, where row data are first loaded to the target storage, and later, typically on-demand, transformed and delivered to end-users, i.e., extract-load-transform (ELT). For instance, an example ELT approach in [100] proposes an ELT architecture that deploys traditional database mechanisms (i.e., hierarchical materialized views) for enabling on-demand data processing of fresher row data previously bulk loaded into a DW.

7.2   Openness

As we have seen, incompleteness in source data (especially in the today's Web oriented environments, see Scenario 2 in Sect. 3), brings several challenges to data-intensive flows. In addition, when integrating and delivering the data at the target, due to possibly incomplete (or non-finite) data, the choice between two main assumptions should be made, i.e., closed world assumption (CWA) or open world assumption (OWA). This choice depends on different characteristics of both, the considered data sources and the expected target. For the systems, like in-house databases or traditional data warehouse systems, where the completeness of data can be assumed, CWA is preferable since in general we do not anticipate discovering additional data (i.e., Low openness). On the other side, when we assume incomplete data at the sources of analysis, we can either follow CWA and create a single finite answer from incomplete data (e.g., by means of data cleaning process), or OWA which would in general allow multiple and possibly an infinite number of interpretations of the answer at the target, by considering also dynamically added data to the analysis [2] (i.e., High openness).

Example. In our example scenarios, in the *traditional BI* setting (Scenario 1), the analysis of the revenue share depends solely on the data about the item sales, currently transferred to DW by means of the ETL process depicted in Fig. 1. On the other hand, the *next generation BI* setting in Scenario 2, should assume a more open environment, where at each moment depending on the end user needs (e.g., following trends and opinions about items as in Examples 2.1 and 2.2) the system must be able to dynamically discover the sources from which such an information can be extracted (e.g., reviews in forums). □

Similarly, [71] revisits two main paradigms for the Semantic Web: (1) *Datalog*, that follows the CWA, and (2) *Classical* (standard logics) paradigm that follows OWA. An important conclusion of this work is that the *Datalog* paradigm as well as CWA is more suitable for highly structured environments in which we can ensure completeness of the data, while the *Classical* paradigm and OWA provide more advantages in loosely coupled environments, where the analysis should not only be limited to the existing (i.e., "in-house") data.

Moreover, coinciding arguments are found in the fields of *data integration* and *data exchange*. Intuitively, CWA (i.e., Low openness) is more suitable assumption in data exchange, where query answering must rely solely on data transferred from source to target using defined schema mappings and not on the data that can be added later [61]. Conversely, more open scenario is typically expected in data integration systems [20], where additional data can be explored on-the-fly and added to the analysis [2] (i.e., High openness). However, the high complexity of query answering under the OWA [71], raises an additional challenge to the latency of data-intensive flows, which is critical in *next generation BI* systems.

7.3 Discussion

The outcome of studying the data delivery stage of data-intensive flows can be seen in Table 3. We observed that the same principles for data delivery (i.e., levels of the studied dimensions in Table 3) are followed in approaches of traditional data exchange [27] and DW settings [38] (i.e., *ETL*). At the same time, we also noticed that the similar principles of the data integration setting [20,59] are envisioned for *next generation BI* systems in some of the studied approachs [1,2] (i.e., *ETO*), while others propose a mixed approach [16] (i.e., *ETL & ETO*).

Table 3. Classification of data delivery appraoches

Data delivery				
ETL vs. ETO	Approaches	TRL	Dimensions	
	AUTHORS, YEAR, [*NAME*,] REFERENCE		*interac.*	*open.*
ETL	Golfarelli & Rizzi, 2009, [38]	2	Low	Low
	Fagin et al., 2005, *Data Exchange*, [27]			
	Libkin, 2006, [61]			
	Theodoratos & Sellis, 1999, [90]			
ETL & ETO	Dayal et al., 2009, [16]	1	Medium	Low
	Qu & Dessloch, 2014, [74]	3		
	Waas et al., 2013, *ELT*, [100]			
	Giacomo et al., 2007, [34]	2	Medium	Medium
ETO	Cugola & Margara, *CEP*, 2012, [14]	1	High	Medium
	Abelló et al., 2013, *Fusion Cubes*, [1]	1	High	High
	Abelló et al., 2015, [2]			
	Lenzerini, 2002, *Data Integration*, [59]	2		
	Halevy, 2001, [42]			
	Ullman, 1997, [94]			
	Doan et al., 2012, *Data Integration* [20]			
	Garcia-Molina et al., 1997, *TSIMMIS*, [32]	3		
	Kirk et al., 1995, *Information Manifold*, [54]			

Such observations strongly indicated that the underpinnings for building a system for managing data-intensive flows in the *next generation BI* setting should be searched in the theoretical field of data integration.

Such a trend has been indeed followed in some of the recent approaches. For example, the idea of creating a unified view over relevant data sources (i.e., the main principle of the data integration setting), is revisited by some approaches by creating a common domain vocabulary and integrating it with existing data sources (e.g., [79,88]). There, the use of a domain ontology to reconcile the languages of business and IT worlds when building a BI system has been proposed. In [79], an ontology is used in combination with schema mappings to automatically generate ETL pipelines to satisfy information requirements previously expressed in terms of an ontology by an end user. The approach works with a predefined set of data sources which is, as suggested, suitable for building a DW system, but as data in today's BI settings are coming from disparate and external data sources, the challenge of capturing their semantics under a common vocabulary brings additional challenges. To this end, Semantic Web technologies are discussed (e.g., [2]) as a solution both for capturing the semantics and further interactive exploration of data, facilitated by the automatic reasoning mechanisms.

8   Optimization of Data-Intensive Flows

Optimizing the execution of data-intensive flows, is a necessity, especially taking into account the *next generation BI* setting that often requires the "right-time" delivery of information.

8.1   Optimization Input

The problem of data flow optimization has been considered from early years of databases from different perspectives, where each of these perspectives may affect different parts of a data flow.

- **Data.** The optimization of a data flow execution can be achieved by transforming the structure of the original dataset (e.g., by means of data partitioning [52]). However, notice that simply transforming a dataset would not achieve a better performance, unless the execution model of a data flow is able to exploit such a transformation (e.g., distributed or parallel data processing [18]).
- **Data flow.** The most typical case considers the optimization of data flow execution by changing the way data are processed, while ensuring the equivalent semantics of the resulting dataset. Such techniques stem from the early years of databases, where minimizing data delivery time by changing the order and selecting the most optimal algorithm for operations applied over input data has been studied under the name of *query optimization* [48]. Moving to the DW environment, which assumes more complex data transformations than the ones in relational algebra, has opened a new field of study dealing with optimizing *ETL* processes (e.g., [83]). In fact, similar principles to those introduced in query optimization (i.e., generating semantically equivalent execution plans for a query by reordering operations, and then finding a plan with a minimal cost) have been applied in [83] and extended to the context of *ETL* flows.

  Another work [44,76] has based operation reordering (i.e., plan rewrites) on automatically discovering a set of extensible operation properties rather than relying solely on algebraic specifications, in order to enable reordering of complex ("black-box") operators. While low data latency is desirable for *ETL* processes, due to limited time windows dedicated to the DW refreshment processes, in the *next generation BI* setting, having data-intensive flows with close to zero latency is a must. Other techniques include: choosing the optimal implementation for the flow operations [93], selecting the optimal execution engine for executing a data flow [56,85], data flow fragmentation and pipelining [52,86].
- **Multi-flow.** In other scenarios, especially in the case of shared execution resources, the optimization goal may suggest optimizing the overall execution of a set of data flows, rather than only the execution of an individual flow. Approaches that deal with this problem fall in two categories. On the one hand, some approaches assume having a detailed knowledge of included data

flows and thus try to exploit it and optimize the overall execution, by means of finding shared parts of data workloads and reusing common execution and data [35,49,80]. Other approaches however assume only a high level knowledge of included data flows (e.g., input data size, execution time, high-level flow complexity, time constraints for the flow execution; [73]). In such cases, the optimization of data flows proceeds by selecting the best scheduling for the execution of data-intensive flows, while the further optimization of individual data flows is left to an engine-specific optimizer [86].

8.2   Dynamicity

While the challenges due to the higher complexity of data transformation has been largely addressed [44,83], proposed cost-based techniques often require certain statistics metadata available for a given data flow in order to find the optimal configuration. However, this is typically not the case and gathering and managing such statistics is not an easy task [41]. [41] proposes a statistics collection framework, by defining a set of necessary statistics, as well as gathering methods. However, this approach although powerful assumes a case of ETL process flow, where data flows are typically designed and optimized in advance (i.e., at design time), while the statistics gathering depends on the previous execution of the same ETL process.

Notice that the majority of the optimization approaches discussed in the previous subsection also assume a static case (see Table 4), where data flows are optimized once, at design time, and then executed many times. The exception to this are approaches that besides statically optimizing a data flow, also provide dynamic optimization of data flow executions in terms of runtime scheduling (i.e., [52,73,86]).

Some optimization approaches however propose on-the-fly gathering of statistics, more suitable for the next generation data flow setting, and applying data flow optimization steps at runtime. The approach in [17] proposes performing micro-benchmarks for building models to estimate the costs of operations using different implementations or executing them on different engines. They show how to deal both with the conventional (relational algebra) operators as well as with complex data transformations typical for the next generation data flows (e.g., sentiment or text analysis). The importance of using more accurate statistics for optimizing data flows in dynamic, cloud-scale environments has been also discussed in [9]. To deal with uncertainty when optimizing running data flows they propose an approach that continuously monitors the execution of data flows at runtime, gathers statistics, and re-optimizes data flows on-the-fly to achieve better performance. The focus here is however on the distributed computational model, where execution times are often higher than in the centralized systems due to necessary synchronization costs, thus the re-optimization overheads are typically considered as negligible.

Table 4. Classification of data flow optimization approaches

Data flow optimization				
ETL vs.ETO	**Approaches**	**TRL**	**Dimensions**	
	AUTHORS, YEAR, [*NAME*,] REFERENCE		*input*	*dynamicity*
ETL	Simitsis et al., 2005, [83]	3	Data flow	Design time
	Hueske et al., 2012, [44]			
	Rheinlnder et al., 2015, *SOFA*, [76]			
	Tziovara et al., 2007, [93]			
	Simitsis et al., 2005, [85]			
	Kougka et al., 2015, [56]			
	Halasipuram et al., 2014, [41]			
	Giannikis et al., 2014, *SharedDB*, [35]	3	Multi-flow	Design time
	Jovanovic et al., 2016, *CoAl*, [49]			
ETO	Karagiannis et al., 2013, [52]	3	Data & Data flow	Runtime
	Dayal et al., 2011, [17]	3	Data flow	Runtime
	Bruno et al., 2013, [9]			
	Jarke & Koch, 1984, [48]	2		
	Simitsis et al., 2013, *HFMS*, [86]	?	Multi-flow	Runtime
	Polo et al., 2014, [7?]	3		

8.3   Discussion

In Table 4, we summarize the outcome of studying data flow optimization approaches in this section. It is easy to see that a static (design time) optimization of data flows has been largely studied in academia. While most approaches worked on the problem of extending traditional query optimization techniques [48] to support more complex data flow operations [44,83], they typically overlook the importance of having the needed statistics of input data and data flow operations to perform cost-based data flow optimization. Such design time optimization approaches require higher overhear and are hence mostly applicable to the traditional BI settings (i.e., *ETL*). Some of the recent approaches insist on the importance of having accurate statistics for creating an optimal execution of a data flow, both for design time [41] and runtime scenarios [9]. Still, the challenges for efficiently gathering and exploiting such statistics metadata for optimizing data-intensive flows remain due to the required close to zero overhead of an optimization process and the "right-time" data delivery demands in the *next generation BI* settings (i.e., *ETO*). To this end, the existing algorithms proposed for efficiently capturing the approximate summaries out of massive data streams [60], should be reconsidered here and adopted for gathering approximate statistics for data-intensive flows over large input datasets.

9   Overall Discussion

Finally, in this section, we summarize the main observations made from the results of our study and propose the high level architecture for managing the lifecycle of data-intensive flows in the *next generation BI* setting. We also give further directions for the topics that require special attention of the research community when studying data-intensive flows.

We have observed some general trends in studying the fields related to data-intensive flows. We focused on the fields of data exchange, data integration, as well as ETL, and ETO. The need for managing heterogeneous data has appeared ever since the database systems start being more broadly used (e.g., federated databases in the 80's [43]). Besides, even though the system in [82] from the 70's is argued to be the first approach that followed the principles of data exchange, the data exchange problem has not been formally defined until the early 00's [26]. Likewise, the problem of data integration is studied from the 90's [32,94], while the strong theoretical overview of the field is given in the early 00's [59]. Along with these theoretical works, the concept of the traditional DW setting was defined by Bill Inmon in the early 90's [46]. ETL, as a separate and rather complex process, however, appeared in the late 90's and the early 00's to replace simple refreshment processes for a DW modeled as a set of materialized views [96]. We can, however, notice the disparity among the trends of studying these fields in the past, showing that they have focused on solving isolated issues.

In the recent years, business environments became more complex, dynamic and interconnected, hence more interactive analytic systems to support daily decision making upon the combination of various (external or internal) data sources, have become a necessity. Moreover, as discussed throughout this paper, today's BI environments require efficiently combining these individual solutions for the problem at hand. To this end, in this paper, we have given a unified view of data-intensive flows, focusing on the challenges that *next generation BI* setting has brought. Currently, even though many works under different names (i.e., from different perspectives) have envisioned and/or proposed conceptual frameworks for *next generation BI* ecosystems (e.g., [1,6,15,17,22,63]), we still lack an end-to-end solution for managing the complete lifecycle of data-intensive flows. Going back to Tables 1, 2 and 3, we can observe a certain overlapping of levels of different dimensions between the theoretical problem of *data exchange* and data warehousing approaches (i.e., *ETL*), as well as between *data integration* and data-intensive flows in the *next generation BI* setting (i.e., *ETO*).

9.1 Architecture for Managing the Lifecycle of Data-Intensive Flows in Next Generation BI Systems

We additionally observed in Tables 1, 2 and 3 that the majority of works supporting the idea of the *next generation BI* setting in fact belong to level 1 of technical readiness ($TRL = 1$), as they are mostly visionary works that analyze the challenges of the *next generation BI* setting from different perspectives. However, we still lack a complete picture of all the aspects of data-intensive flows in this new setting, and to this end, we envision here an architecture of a system for managing data-intensive flows (Fig. 6).

The proposed architecture depicts at high level the main outcome of our study. It points out the main processing steps which need to be considered during the lifecycle of a data-intensive flow. Moreover, the architecture captures in a holistic way the complete lifecycle of data-intensive flows, and as such, it can be seen as a roadmap for both academia and industry toward building data-intensive flows in *next generation BI* systems. In what follows, we discuss in more detail different architectural modules, and if available, we point out example approaches that tackle the challenges of such modules.

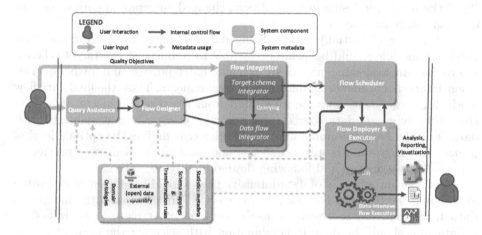

Fig. 6. Architecture for managing the lifecycle of data-intensive flows

We start with the *Query Assistance* module that should provide an intuitive interface to end users when expressing their information requirements. On the one hand, it should raise the usability of the BI system, for a broader set of business users. This module should provide a business-oriented view over the included data sources (e.g., by means of domain ontologies like in [5,50,79]). On the other hand, the *Query Assistance* module should also facilitate the low *coupledness* of data sources and be able to efficiently connect to a plethora of external data source repositories. These, preferably semantically enriched data sources (e.g., linked (open) data; [7]), should supplement user analysis with context-aware data and thus raise the *openness* of the delivered results (see Sect. 7).

Previously expressed information requirements further need to be automatically translated in an appropriate data flow (i.e., the *Flow Designer* module) that will satisfy such information requirements. *Flow Designer* must provide robustness for such loosely coupled systems in dealing with data sources with low (non-guaranteed) *accessibility*. Furthermore, to support the high *automation* of the *Flow Designer* module, the system should first revisit the existing approaches for automatic schema mapping creation (e.g., [24]). Obviously, these approaches must be extended with more complex data transformations. First, supporting low *structuredness* and extracting useful data from unstructured data sources on-the-fly should be largely supported, as dynamic systems cannot always rely on having a batched preprocessing step doing so. Furthermore, more complex data analysis (e.g., machine learning algorithms) should be supported in these data flows. Here, we can benefit from exploiting different data and flow characteristics (e.g., by revisiting the previously studied field of intelligent assistants for data analysis [81]). Lastly, the *Flow Designer* module should automatically accommodate the flow to ensure the required level of data quality and output data *constraintness* (typically in contradiction with required data latency) [91]. Importantly, such a design process must be iterative to support high *malleability* of the data flow design in front of new, changed, or removed data sources or information requirements.

In the case of partially materializing data, as suggested in [15], the *Flow Designer* module should be aware or be able to reconstruct the target schema, where data are loaded in a previously executed batch process, and further queried when *interactively* answering information requirements. Thus, the final data flow ready for deployment must be integrated from the combination of data flows that directly access data sources and querying previously materialized target data (i.e., the *Flow Integrator* module). Notice that finding the optimal level of partial materialization is still a challenge and must be decided using previously collected flow statistics and following desired quality objectives.

Next, the optimization of data-intensive flows should be efficiently supported at different levels of the flow lifecyle (see Sect. 8. Initially, when the integrated data flow is created to answer a user's requirement at hand, optimization of a data flow should be done in combination with selecting the optimal partial materialization of data [74]. Furthermore, having multiple data-intensive flows answering different requirements of end-users waiting for execution, the system requires an optimal schedule for running these data flows over the shared computational resources (e.g., shared, multi-tenant cluster), i.e., *Flow Scheduler* module. Lastly, the automatic optimization means must be also provided when deploying data flows, for selecting an optimal execution engine (e.g., [56,85]), as well as for providing the lower level, engine-specific, optimization of a data flow (i.e., the *Flow Deployer* module).

From Fig. 6 we can observe that the *automation* of the design and optimization of data-intensive flows, as well as the query assistance, must be largely facilitated by means of different metadata artifacts (i.e., schema mappings, domain ontology, flow and statistics). Indeed, the use of metadata for automating the design of the next generation data warehouse systems (DW 2.0) has been previously discussed [47], while recently the main challenges of matadata in the analytical process of the *next generation BI* systems have been studied in [95].

Finally, as an important remark, we want to draw a parallelism of the envisioned architecture depicted in Fig. 6, and the traditional architecture of centralized database management systems (DBMS). First, using declarative (SQL) queries in a DBMS, end users pose their analytical needs to the system. While based on the traditional database theory, the *semantic optimizer* is responsible for transforming a user query into an equivalent one with a lower cost, in *next generation BI* systems, user queries need to be additionally transformed and enriched to access external data by means of data exploration processes (i.e.,

Query Assistance; [2]). Furthermore, similarly to the *syntactic optimizer* in the traditional DBMSs, *Flow Designer* needs to translate an information requirement to a sequence of operations (i.e., *syntactic tree*), which represents a logical plan of a data-intensive flow execution. The execution plan should be typically optimized for an individual execution. However, in *next generation BI* systems, a data-intensive flow could also be integrated with other data flows for an optimized *multi-flow* execution (i.e., *Flow Integrator*). This conceptually resembles the well-known problem of multi-query optimization [80], but inevitably brings new challenges considering the complexity of data flow operations, which cannot always be presented using algebraic specifications [44]. Moreover, the *Flow Integrator* module should also transform input execution plan and optimize it considering partial materialization of data, similarly to the *query rewriting* techniques for answering queries using materialized views [42]. Following the traditional DBMS architecture, an integrated execution plan is then optimally scheduled for execution, together with the rest of the data flows in the system (i.e., *Flow Scheduler*). Lastly, the logical data flow is translated into the code of a selected execution engine (e.g., [51,56]), *physically optimized* considering available access structures, and finally deployed for *execution* (i.e., *Flow Deployer & Executor*). Similarly to the concept of the database catalog, throughout the lifecycle of a data intensive flow, different metadata artifacts need to be available (e.g., schema mappings, transformation rules, statistics metadata; see Fig. 6) to lead the automatic design and optimization of a data flow.

The parallelism drawn above finally confirms us that the underpinnings of data-intensive flows in *next generation BI* systems should be analyzed in the frame of the traditional DB theory field. Nevertheless, as we showed through our study, the inherent complexity of today's business environments (e.g., data heterogeneity, high complexity of data processing) must be additionally addressed, and comprehensively tackled to provide end-to-end solutions for managing the complete lifecycle of data-intensive flows.

10 Conclusions

In this paper, we studied data-intensive flows, focusing on the challenges of the next-generation BI setting. We analyzed the foundational work of database theory tackling heterogeneity and interoperability (i.e., data exchange and data integration), as well as the recent approaches both in the context of DW and *next generation BI* systems.

We first identified the main characteristics of data-intensive flows, which built the dimensions of our study setting, and further studied the current approaches in the frame of these dimensions and determined the level the studied approaches attain in each of them.

As the main outcome of this study, we outline an architecture for managing the complexity of data-intensive flows in the *next generation BI* setting. We discuss in particular different components that such an architecture should realize, as well as the processes that the data-intensive flow lifecycle should carry out.

Finally, within the components of the envisioned architecture, we point out the main remaining challenges that the *next generation BI* setting brings to managing data-intensive flows, and which require special attention from both academia and industry.

Acknowledgements. This work has been partially supported by the Secreteria d'Universitats i Recerca de la Generalitat de Catalunya under 2014 SGR 1534, and by the Spanish Ministry of Education grant FPU12/04915.

References

1. Abelló, A., Darmont, J., Etcheverry, L., Golfarelli, M., Mazón, J.N., Naumann, F., Pedersen, T.B., Rizzi, S., Trujillo, J., Vassiliadis, P., Vossen, G.: Fusion cubes: towards self-service business intelligence. IJDWM **9**(2), 66–88 (2013)
2. Abelló, A., Romero, O., Pedersen, T.B., Llavori, R.B., Nebot, V., Cabo, M.J.A., Simitsis, A.: Using semantic web technologies for exploratory OLAP: a survey. IEEE Trans. Knowl. Data Eng. **27**(2), 571–588 (2015)
3. Akkaoui, Z., Mazón, J.-N., Vaisman, A., Zimányi, E.: BPMN-based conceptual modeling of ETL processes. In: Cuzzocrea, A., Dayal, U. (eds.) DaWaK 2012. LNCS, vol. 7448, pp. 1–14. Springer, Heidelberg (2012). doi:10.1007/978-3-642-32584-7_1
4. Akkaoui, Z.E., Zimányi, E., Mazón, J.N., Trujillo, J.: A BPMN-based design and maintenance framework for ETL processes. IJDWM **9**(3), 46–72 (2013)
5. Bellatreche, L., Khouri, S., Berkani, N.: Semantic data warehouse design: from ETL to deployment à la carte. In: Meng, W., Feng, L., Bressan, S., Winiwarter, W., Song, W. (eds.) DASFAA 2013. LNCS, vol. 7826, pp. 64–83. Springer, Heidelberg (2013). doi:10.1007/978-3-642-37450-0_5
6. Berthold, H., Rösch, P., Zöller, S., Wortmann, F., Carenini, A., Campbell, S., Bisson, P., Strohmaier, F.: An architecture for ad-hoc and collaborative business intelligence. In: EDBT/ICDT Workshops (2010)
7. Bizer, C., Heath, T., Berners-Lee, T.: Linked data - the story so far. Int. J. Semantic Web Inf. Syst. **5**(3), 1–22 (2009)
8. Bruckner, R.M., List, B., Schiefer, J.: Striving towards near real-time data integration for data warehouses. In: Kambayashi, Y., Winiwarter, W., Arikawa, M. (eds.) DaWaK 2002. LNCS, vol. 2454, pp. 317–326. Springer, Heidelberg (2002). doi:10.1007/3-540-46145-0_31
9. Bruno, N., Jain, S., Zhou, J.: Continuous cloud-scale query optimization and processing. PVLDB **6**(11), 961–972 (2013)
10. Buneman, P., Davidson, S., Fernandez, M., Suciu, D.: Adding structure to unstructured data. In: Afrati, F., Kolaitis, P. (eds.) ICDT 1997. LNCS, vol. 1186, pp. 336–350. Springer, Heidelberg (1997). doi:10.1007/3-540-62222-5_55
11. Calì, A., Calvanese, D., De Giacomo, G., Lenzerini, M.: Data integration under integrity constraints. Inf. Syst. **29**(2), 147–163 (2004)
12. Calì, A., Lembo, D., Rosati, R.: Query rewriting and answering under constraints in data integration systems. In: IJCAI, pp. 16–21 (2003)
13. Chen, C.L.P., Zhang, C.: Data-intensive applications, challenges, techniques and technologies: a survey on big data. Inf. Sci. **275**, 314–347 (2014)
14. Cugola, G., Margara, A.: Processing flows of information: from data stream to complex event processing. ACM Comput. Surv. **44**(3), 15 (2012)

15. Dayal, U., Castellanos, M., Simitsis, A., Wilkinson, K.: Data integration flows for business intelligence. In: EDBT, pp. 1–11 (2009)
16. Dayal, U., Kuno, H.A., Wiener, J.L., Wilkinson, K., Ganapathi, A., Krompass, S.: Managing operational business intelligence workloads. Operating Syst. Rev. **43**(1), 92–98 (2009)
17. Dayal, U., Wilkinson, K., Simitsis, A., Castellanos, M., Paz, L.: Optimization of analytic data flows for next generation business intelligence applications. In: Nambiar, R., Poess, M. (eds.) TPCTC 2011. LNCS, vol. 7144, pp. 46–66. Springer, Heidelberg (2012). doi:10.1007/978-3-642-32627-1_4
18. Dean, J., Ghemawat, S.: Mapreduce: simplified data processing on large clusters. Commun. ACM **51**(1), 107–113 (2008)
19. Dessloch, S., Hernández, M.A., Wisnesky, R., Radwan, A., Zhou, J.: Orchid: integrating schema mapping and ETL. In: ICDE, pp. 1307–1316. IEEE (2008)
20. Doan, A., Halevy, A.Y., Ives, Z.G.: Principles of Data Integration. Morgan Kaufmann, Waltham (2012)
21. Dong, X.L., Halevy, A.Y., Yu, C.: Data integration with uncertainty. VLDB J. **18**(2), 469–500 (2009)
22. Eckerson, W.W.: Best practices in operational BI. Bus. Intell. J. **12**(3), 7–9 (2007)
23. European Commission: G. technology readiness levels (TRL) (2014)
24. Fagin, R., Haas, L.M., Hernández, M., Miller, R.J., Popa, L., Velegrakis, Y.: Clio: schema mapping creation and data exchange. In: Borgida, A.T., Chaudhri, V.K., Giorgini, P., Yu, E.S. (eds.) Conceptual Modeling: Foundations and Applications. LNCS, vol. 5600, pp. 198–236. Springer, Heidelberg (2009). doi:10.1007/978-3-642-02463-4_12
25. Fagin, R., Kimelfeld, B., Kolaitis, P.G.: Probabilistic data exchange. J. ACM (JACM) **58**(4), 15 (2011)
26. Fagin, R., Kolaitis, P.G., Miller, R.J., Popa, L.: Data exchange: semantics and query answering. In: Calvanese, D., Lenzerini, M., Motwani, R. (eds.) ICDT 2003. LNCS, vol. 2572, pp. 207–224. Springer, Heidelberg (2003). doi:10.1007/3-540-36285-1_14
27. Fagin, R., Kolaitis, P.G., Miller, R.J., Popa, L.: Data exchange: semantics and query answering. Theor. Comput. Sci. **336**(1), 89–124 (2005)
28. Feldman, R., Sanger, J.: The Text Mining Handbook: Advanced Approaches in Analyzing Unstructured Data. Cambridge University Press, New York (2007)
29. Ferrara, E., Meo, P.D., Fiumara, G., Baumgartner, R.: Web data extraction, applications and techniques: a survey. Knowl. Based Syst. **70**, 301–323 (2014)
30. Friedman, M., Levy, A.Y., Millstein, T.D.: Navigational Plans for Data Integration. In: Intelligent Information Integration (1999)
31. García, S., Romero, O.: Ravents: R.: DSS from an RE perspective: a systematic mapping. J. Syst. Softw. **117**, 488–507 (2016)
32. Garcia-Molina, H., Papakonstantinou, Y., Quass, D., Rajaraman, A., Sagiv, Y., Ullman, J.D., Vassalos, V., Widom, J.: The TSIMMIS approach to mediation: data models and languages. J. Intell. Inf. Syst. **8**(2), 117–132 (1997)
33. Ghazal, A., Rabl, T., Hu, M., Raab, F., Poess, M., Crolotte, A., Jacobsen, H.A.: BigBench: towards an industry standard benchmark for big data analytics. In: SIGMOD Conference, pp. 1197–1208 (2013)
34. Giacomo, G.D., Lembo, D., Lenzerini, M., Rosati, R.: On reconciling data exchange, data integration, and peer data management. In: PODS, pp. 133–142 (2007)
35. Giannikis, G., Makreshanski, D., Alonso, G., Kossmann, D.: Shared workload optimization. PVLDB **7**(6), 429–440 (2014)

36. Giorgini, P., Rizzi, S., Garzetti, M.: Grand: a goal-oriented approach to require-
    ment analysis in data warehouses. DSS **45**(1), 4–21 (2008)
37. Golfarelli, M., Lechtenbörger, J., Rizzi, S., Vossen, G.: Schema versioning in
    data warehouses: enabling cross-version querying via schema augmentation. Data
    Knowl. Eng. **59**(2), 435–459 (2006)
38. Golfarelli, M., Rizzi, S.: Data Warehouse Design. Modern Principles and Method-
    ologies. McGraw-Hill (2009)
39. Golfarelli, M., Rizzi, S., Cella, I.: Beyond data warehousing: what's next in busi-
    ness intelligence?. In: DOLAP, pp. 1–6 (2004)
40. Haas, L.: Beauty and the beast: the theory and practice of information integration.
    In: Schwentick, T., Suciu, D. (eds.) ICDT 2007. LNCS, vol. 4353, pp. 28–43.
    Springer, Heidelberg (2006). doi:10.1007/11965893_3
41. Halasipuram, R., Deshpande, P.M., Padmanabhan, S.: Determining essential sta-
    tistics for cost based optimization of an ETL workflow. In: EDBT, pp. 307–318
    (2014)
42. Halevy, A.Y.: Answering queries using views: a survey. VLDB J. **10**(4), 270–294
    (2001)
43. Heimbigner, D., McLeod, D.: A federated architecture for information manage-
    ment. ACM Trans. Inf. Syst. **3**(3), 253–278 (1985)
44. Hueske, F., Peters, M., Sax, M., Rheinländer, A., Bergmann, R., Krettek, A.,
    Tzoumas, K.: Opening the black boxes in data flow optimization. PVLDB **5**(11),
    1256–1267 (2012)
45. Zikopoulos, P., Eaton, C.: Understanding Big Data: Analytics for Enterprise Class
    Hadoop and Streaming Data. McGraw-Hill Osborne Media, 1st edn. (2011)
46. Inmon, W.H.: Building the Data Warehouse. John Wiley & Sons, Inc. (1992)
47. Inmon, W.H., Strauss, D., Neushloss, G.: DW 2.0: The architecture for the next
    generation of data warehousing: The architecture for the next generation of data
    warehousing. Morgan Kaufmann (2010)
48. Jarke, M., Koch, J.: Query optimization in database systems. ACM Comput.
    Surv. **16**(2), 111–152 (1984)
49. Jovanovic, P., Romero, O., Simitsis, A., Abelló, A.: Incremental consolidation
    of data-intensive multi-flows. IEEE Trans. Knowl. Data Eng. **28**(5), 1203–1216
    (2016)
50. Jovanovic, P., Romero, O., Simitsis, A., Abelló, A., Candón, H., Nadal, S.: Quarry:
    Digging up the gems of your data treasury. In: EDBT, pp. 549–552 (2015)
51. Jovanovic, P., Simitsis, A., Wilkinson, K.: Engine independence for logical analytic
    flows. In: ICDE, pp. 1060–1071 (2014)
52. Karagiannis, A., Vassiliadis, P., Simitsis, A.: Scheduling strategies for efficient
    ETL execution. Inf. Syst. **38**(6), 927–945 (2013)
53. Kimball, R., Caserta, J.: The Data Warehouse ETL Toolkit. John Wiley & Sons
    (2004)
54. Kirk, T., Levy, A.Y., Sagiv, Y., Srivastava, D.: Others: The information manifold.
    In: Proceedings of the AAAI 1995 Spring Symposium on Information Gathering
    from Heterogeneous, Distributed Enviroments, vol. 7, pp. 85–91 (1995)
55. Kolaitis, P.G.: Schema mappings, data exchange, and metadata management. In:
    PODS, pp. 61–75 (2005)
56. Kougka, G., Gounaris, A., Tsichlas, K.: Practical algorithms for execution engine
    selection in data flows. Future Gener. Comput. Syst. **45**, 133–148 (2015)
57. Labio, W., Garcia-Molina, H.: Efficient snapshot differential algorithms for data
    warehousing. In: VLDB, pp. 63–74 (1996)

58. Laender, A.H.F., Ribeiro-Neto, B.A., da Silva, A.S., Teixeira, J.S.: A brief survey of web data extraction tools. SIGMOD Rec. **31**(2), 84–93 (2002)
59. Lenzerini, M.: Data integration: a theoretical perspective. In: PODS, pp. 233–246. ACM (2002)
60. Leskovec, J., Rajaraman, A., Ullman, J.D.: Mining of Massive Datasets. Cambridge University Press, New York (2014)
61. Libkin, L.: Data exchange and incomplete information. In: PODS, pp. 60–69 (2006)
62. Lindsay, B.G., Haas, L.M., Mohan, C., Pirahesh, H., Wilms, P.F.: A snapshot differential refresh algorithm. In: SIGMOD Conference, pp. 53–60 (1986)
63. Löser, A., Hueske, F., Markl, V.: Situational business intelligence. In: Castellanos, M., Dayal, U., Sellis, T. (eds.) BIRTE 2008. LNBIP, vol. 27, pp. 1–11. Springer, Heidelberg (2009). doi:10.1007/978-3-642-03422-0_1
64. Mazón, J.N., Lechtenbörger, J., Trujillo, J.: A survey on summarizability issues in multidimensional modeling. Data Knowl. Eng. **68**(12), 1452–1469 (2009)
65. McBrien, P., Poulovassilis, A.: Data integration by bi-directional schema transformation rules. In: ICDE, pp. 227–238 (2003)
66. McDonald, K., Wilmsmeier, A., Dixon, D.C., Inmon, W.: Mastering the SAP Business Information Warehouse. John Wiley & Sons (2002)
67. Morzy, T., Wrembel, R.: On querying versions of multiversion data warehouse. In: DOLAP, pp. 92–101 (2004)
68. Muñoz, L., Mazón, J.N., Trujillo, J.: Automatic generation of ETL processes from conceptual models. In: DOLAP, pp. 33–40 (2009)
69. Ong, K.W., Papakonstantinou, Y., Vernoux, R.: The SQL++ semi-structured data model and query language: a capabilities survey of sql-on-hadoop, nosql and newsql databases. CoRR abs/1405.3631 (2014). http://arxiv.org/abs/1405.3631
70. Papastefanatos, G., Vassiliadis, P., Simitsis, A., Vassiliou, Y.: Policy-Regulated Management of ETL Evolution. In: Spaccapietra, S., Zimányi, E., Song, I.-Y. (eds.) Journal on Data Semantics XIII. LNCS, vol. 5530, pp. 147–177. Springer, Heidelberg (2009). doi:10.1007/978-3-642-03098-7_6
71. Patel-Schneider, P.F., Horrocks, I.: Position paper: a comparison of two modelling paradigms in the Semantic Web. In: WWW, pp. 3–12 (2006)
72. Pohl, K.: Requirements Engineering - Fundamentals, Principles, and Techniques. Springer, Heidelberg (2010)
73. Polo, J., Becerra, Y., Carrera, D., Torres, J., Ayguadé, E., Steinder, M.: Adaptive MapReduce scheduling in shared environments. In: IEEE/ACM CCGrid, pp. 61–70 (2014)
74. Qu, W., Dessloch, S.: A real-time materialized view approach for analytic flows in hybrid cloud environments. Datenbank-Spektrum **14**(2), 97–106 (2014)
75. Rahm, E., Do, H.H.: Data cleaning: problems and current approaches. IEEE Data Eng. Bull. **23**(4), 3–13 (2000)
76. Rheinlnder, A., Heise, A., Hueske, F., Leser, U., Naumann, F.: Sofa: an extensible logical optimizer for UDF-heavy data flows. Inf. Syst. **52**, 96–125 (2015)
77. Romero, O., Abelló, A.: A survey of multidimensional modeling methodologies. IJDWM **5**(2), 1–23 (2009)
78. Romero, O., Abelló, A.: Open access semantic aware business intelligence. In: Zimányi, E. (ed.) eBISS 2013. LNBIP, vol. 172, pp. 121–149. Springer, Heidelberg (2014). doi:10.1007/978-3-319-05461-2_4

79. Romero, O., Simitsis, A., Abelló, A.: *GEM*: requirement-driven generation of ETL and multidimensional conceptual designs. In: Cuzzocrea, A., Dayal, U. (eds.) DaWaK 2011. LNCS, vol. 6862, pp. 80–95. Springer, Heidelberg (2011). doi:10. 1007/978-3-642-23544-3_7

80. Roy, P., Sudarshan, S.: Multi-query optimization. In: Encyclopedia of Database Systems, pp. 1849–1852. Springer, US (2009)

81. Serban, F., Vanschoren, J., Kietz, J., Bernstein, A.: A survey of intelligent assistants for data analysis. ACM Comput. Surv. **45**(3), 31 (2013)

82. Shu, N.C., Housel, B.C., Taylor, R.W., Ghosh, S.P., Lum, V.Y.: EXPRESS: a data extraction, processing, amd REStructuring system. ACM Trans. Database Syst. **2**(2), 134–174 (1977)

83. Simitsis, A., Vassiliadis, P., Sellis, T.K.: State-space optimization of ETL work-flows. IEEE Trans. Knowl. Data Eng. **17**(10), 1404–1419 (2005)

84. Simitsis, A., Wilkinson, K., Castellanos, M., Dayal, U.: QoX-driven ETL design: reducing the cost of ETL consulting engagements. In: SIGMOD, pp. 953–960 (2009)

85. Simitsis, A., Wilkinson, K., Castellanos, M., Dayal, U.: Optimizing analytic data flows for multiple execution engines. In: SIGMOD Conference, pp. 829–840 (2012)

86. Simitsis, A., Wilkinson, K., Dayal, U., Hsu, M.: HFMS: managing the lifecycle and complexity of hybrid analytic data flows. In: ICDE, pp. 1174–1185 (2013)

87. Skoutas, D., Simitsis, A.: Designing ETL processes using semantic web technolo-gies. In: DOLAP, pp. 67–74 (2006)

88. Skoutas, D., Simitsis, A.: Ontology-based conceptual design of ETL processes for both structured and semi-structured data. Int. J. Semantic Web Inf. Syst. **3**(4), 1–24 (2007)

89. Strange, K.H.: ETL Was the Key to This Data Warehouse's Success. Gartner Research, CS-15-3143 (2002)

90. Theodoratos, D., Sellis, T.K.: Designing data warehouses. Data Knowl. Eng. **31**(3), 279–301 (1999)

91. Theodorou, V., Abelló, A., Thiele, M., Lehner, W.: POIESIS: a tool for quality-aware ETL process redesign. In: EDBT, pp. 545–548 (2015)

92. Trujillo, J., Luján-Mora, S.: A UML based approach for modeling ETL processes in data warehouses. In: ER, pp. 307–320 (2003)

93. Tziovara, V., Vassiliadis, P., Simitsis, A.: Deciding the physical implementation of ETL workflows. In: DOLAP, pp. 49–56 (2007)

94. Ullman, J.D.: Information integration using logical views. In: Afrati, F., Kolaitis, P. (eds.) ICDT 1997. LNCS, vol. 1186, pp. 19–40. Springer, Heidelberg (1997). doi:10.1007/3-540-62222-5_34

95. Varga, J., Romero, O., Pedersen, T.B., Thomsen, C.: Towards next generation BI systems: the analytical metadata challenge. In: DaWaK, pp. 89–101 (2014)

96. Vassiliadis, P.: A survey of extract-transform-load technology. IJDWM **5**(3), 1–27 (2009)

97. Vassiliadis, P., Simitsis, A.: Near real time ETL. In: New Trends in Data Ware-housing and Data Analysis, pp. 1–31. Springer, US (2009)

98. Vassiliadis, P., Simitsis, A., Georgantas, P., Terrovitis, M., Skiadopoulos, S.: A generic and customizable framework for the design of ETL scenarios. Inf. Syst. **30**(7), 492–525 (2005)

99. Vassiliadis, P., Simitsis, A., Skiadopoulos, S.: Conceptual modeling for ETL processes. In: DOLAP, pp. 14–21 (2002)

100. Waas, F., Wrembel, R., Freudenreich, T., Thiele, M., Koncilia, C., Furtado, P.: On-demand ELT architecture for right-time BI: extending the vision. IJDWM 9(2), 21–38 (2013)
101. Wilkinson, K., Simitsis, A., Castellanos, M., Dayal, U.: Leveraging business process models for ETL design. In: Parsons, J., Saeki, M., Shoval, P., Woo, C., Wand, Y. (eds.) ER 2010. LNCS, vol. 6412, pp. 15–30. Springer, Heidelberg (2010). doi:10.1007/978-3-642-16373-9_2
102. Winter, R., Strauch, B.: A method for demand-driven information requirements analysis in data warehousing projects. In: Proceedings of the HICSS, pp. 1359–1365 (2003)

A Self-Adaptive and Incremental Approach for Data Profiling in the Semantic Web

Kenza Kellou-Menouer[(✉)] and Zoubida Kedad

DAVID Laboratory - University of Versailles Saint-Quentin-en-Yvelines,
Versailles, France
{kenza.menouer,zoubida.kedad}@uvsq.fr

Abstract. The increasing adoption of linked data principles has led to the availability of a huge amount of datasets on the Web. However, the use of these datasets is hindered by the lack of descriptive information about their content. Indeed, interlinking, matching or querying them requires some knowledge about the types and properties they contain.

In this paper, we tackle the problem of describing the content of an RDF dataset by profiling its entities, which consists in discovering the implicit types and providing their description. Each type is described by a profile composed of properties and their probabilities. Our approach relies on a clustering algorithm. It is self-adaptive, as it can automatically detect the most appropriate similarity threshold according to the dataset. Our algorithms generate overlapping clusters, enabling the detection of several types for an entity. As a dataset may evolve, our approach is incremental and can assign a type to a new entity and update the type profiles without browsing the whole dataset. We also present some experimental evaluations to demonstrate the effectiveness of our approach.

Keywords: Similarity threshold · Clustering · Classification · RDF data

1   Introduction

The emergence of the Web of data has led to the availability of a huge number of linked datasets, enabling the development of novel applications. In order to exploit these datasets in a meaningful way, applications and users need some knowledge about them such as the types they contain and their properties. This information is crucial to understand the content of a dataset and it is very useful in many tasks such as creating links between datasets and querying them. Browsing the datasets to understand their content is impossible when their size becomes important. Having a description of the types in the dataset is thus essential for its exploitation.

Linked datasets are subject to constant evolution and may contain noise. In addition, the nature of the languages used to describe data on the Web,

© Springer-Verlag GmbH Germany 2016
A. Hameurlain et al. (Eds.): TLDKS XXIX, LNCS 10120, pp. 108–133, 2016.
DOI: 10.1007/978-3-662-54037-4_4

such as RDF[1](S[2])/OWL[3], do not impose any constraint on the structure of these data: resources of the same type may not have the same properties, and an entity may have several types. Most linked datasets are incomplete with respect to type information, even when linked data are automatically extracted from a controlled source, such as DBpedia [4] which is extracted from Wikipedia.

The goal of our work is to profile an RDF dataset which consists in grouping entities of the same type, even when type declarations are missing or incomplete, and providing a profile for each group. The profile of a group of entities is a property vector where each property is associated to the probability of an entity of this type to have this property.

The requirements for our profiling approach are the followings: (i) the number of types in the dataset is not known, (ii) the dataset may contain noise, (iii) an entity can have several types and (iiii) the dataset evolves with additions of new entities. In our previous work [23], we have used the DBscan clustering algorithm [15] as it is deterministic, it detects noise and it finds clusters of arbitrary shape, which is useful for datasets where entities are described with heterogeneous property sets. In addition, unlike the algorithms based on k-means [20], the number of clusters is not required. However, the similarity threshold parameter must be specified for the clustering. This parameter is very important for the quality of the resulting types. Our first contribution in this paper is an approach for the automatic detection of the similarity threshold.

Another of our requirements that the DBscan algorithm does not meet is that it can not assign several types to an entity because the discovered clusters are disjoint. Fuzzy clustering algorithms such as FCM [6] or EM [13] could be used to assign several types to an entity. However, they require the number of clusters as their grouping criterion is the similarity between an entity and the center of each cluster. Our second contribution is an algorithm for detecting overlapping types and allowing multiple typing of entities. As the dataset evolves and new entities are added, one problem is the determination of the types of an incoming entity. Our third contribution is an incremental profiling approach tackling this issue.

In this paper, we propose a self-adaptive approach for profiling entities of an RDF dataset, which automatically detects the most appropriate similarity threshold according to the dataset. It enables multiple typing of entities and it is incremental in order to cope with the dataset evolutions. The paper is organized as follows. We present our motivations in Sect. 2, which shows some use cases for our approach. We introduce some preliminaries on linked data in Sect. 3 and the use of DBscan for clustering these data in Sect. 4. In Sect. 5, we present our approach for profiling linked data. Our evaluation results are presented in Sect. 6. We discuss the related works in Sect. 7 and finally, Sect. 8 concludes the paper.

---

[1] Resource Description Framework: http://www.w3.org/RDF/.

[2] RDF Schema: http://www.w3.org/TR/rdf-schema/.

[3] Web Ontology Language: http://www.w3.org/OWL/.

2   Motivation

Getting the big picture of a large RDF dataset and characterizing its content is often difficult. Understanding a dataset becomes even more difficult when schema related information is missing. Our approach for profiling the entities of a dataset is a contribution to this task. The goal of profiling is to extract metadata providing information about a dataset. For example, in [7,17], profiling consists in characterizing a dataset by the number of types it contains, the number of instances for each type, the number of instances having a given property, etc. In our context, profiling consists in providing the implicit types of entities as well as a description of these types in terms of their properties and associated occurrence probabilities. The resulting profiles could be useful in various data processing and data management tasks [28], we have listed some of them in the following.

Schema Induction and Data Aggregation. Data profiling helps to determine the adequate schema of a dataset. The discovered types and the associated profiles provided by our approach can also be used to discover the schema of the dataset. Discovering the types is the first step towards building the whole schema, including the semantic and the hierarchical links between these types. The profiles provided by our algorithms can be used to discover these links as proposed in [23]. The generated clusters can also be exploited for more complex RDF graph aggregation as proposed in [26], which requires having instances grouped according to their types. In [25], a SPARQL query-based approach is proposed for semantic Web data profiling to construct the actual schema. However, like Schemex [24], it is based on schema related declarations such as *rdf:type*. Our work could complement these approaches by assigning the untyped entities of the dataset to the correct classes.

Distributed Query Optimization. The goal of this task is to identify the relevant sources for a query and to find optimal execution plans as proposed in [31]. When a query is used over several data sources, the set of profiles of each source could help in decomposing the query and sending the sub-queries to the relevant sources. The probabilities of the properties could be useful in order to optimize the execution plans by ordering the sub-queries according to the selectivity of their criteria. Indeed, our type profiles summarize the set of properties describing the entities of each type which helps to formulate the query. In addition, the probabilities of the properties in the type profiles help to estimate the costs of the execution plans.

Data Integration and Linking. These tasks are often hindered by the lack of information on datasets. Data profiling reveals information on the schema and the semantics of the datasets. Some interlinking tools have been proposed, such as Knofuss[4] or Silk[5], which was used to link Yago [35] to DBpedia [4]. But they require type and property information about the datasets to generate the

---

[4] Knofuss: technologies.kmi.open.ac.uk/knofuss.
[5] Silk: wifo5-03.informatik.uni-mannheim.de/bizer/silk.

appropriate *owl:sameAs* links. These types and properties are described by the type profiles provided by our approach.

Data Quality. Data profiling techniques help to validate datasets against schema properties, and detect irregularities for improving data quality. In our approach, the probabilities in the profiles of the types reflect the gap between the instances and the schema as described in [22], where a set of quality factors is proposed to evaluate this gap.

3   Preliminaries on Linked Data

The context of our work is the Web of data, where linked datasets are published following a set of principles [19] such as using URIs as names for things and including links to other URIs. These datasets are expressed using concepts proposed by W3C [1]. Figure 1 is an example of a linked dataset. The nodes are connected using links representing properties. A node can represent either a class such as "Author" in our example, or an instance linked to its class by the *rdf:type* property, such as "Dumas", or a literal. As we can see in the example, a dataset may include both schema related information, which could be a class defined in the dataset such as "Author", and instance related information, such as the properties of "Dumas". More generally, we define an RDF data graph as follows.

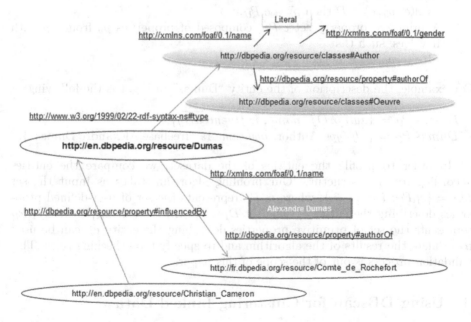

Fig. 1. Example of linked Data.

Definition 1 (RDF Data Graph). Consider the sets R, B, P and L representing resources, blank nodes (anonymous resources), properties and literals respectively. A dataset described in RDF(S)/OWL is defined as a set of triples $D \subseteq (R \cup B) \times P \times (R \cup B \cup L)$. Graphically, the dataset is represented by a labeled directed graph G, where each node is a resource, a blank node or a literal and where each edge from a node e to another node e' labeled with the property p represents the triple (e, p, e') of the dataset D. In such RDF graph, we define an entity as a node corresponding to either a resource or a blank node, that is, any node apart from the ones corresponding to literals.

In our approach, we consider that an entity is described by different kinds of properties. Some of them are part of the RDF(S)/OWL vocabularies, and we will refer to them as primitive properties, and others are user-defined. We distinguish between these two kinds because all the properties should not be used for the same purpose during type profiling. Some predefined properties could be applied to any entity, therefore they should not be considered when evaluating the similarity between entities.

Definition 2 (Entity Description). Given the set of primitive properties P_P and the set of user-defined properties P_U in the dataset D, an entity e is described by:

1. A user-defined property set $e.P_U$ composed of properties p_u from P_U, each one annotated by an arrow indicating its direction, and such that:
   - If $\exists (e, p_u, e') \in D$ then $\overrightarrow{p_u} \in e.P_U$;
   - If $\exists (e', p_u, e) \in D$ then $\overleftarrow{p_u} \in e.P_U$.
2. A primitive property set $e.P_P$, composed of properties p_p from P_P with their values, such that:
   - If $\exists (e, p_p, e') \in D$ then $p_p \in e.P_P$.

For example, the description of the entity "Dumas" in Fig. 1 is the following:

- $Dumas.P_U = \{\overrightarrow{authorOf}, \overrightarrow{name}, \overleftarrow{influencedBy}\}$;
- $Dumas.P_P = \{rdf{:}type$ Author, $owl{:}sameAs$ "freebase:Alexandre Dumas"$\}$.

In order to profile the entities of the dataset, we compare the entities according to their structure. Our profiling algorithm takes as input the set $D_U = \{e_i.P_U\colon i = 1, ...n\}$, where $e_i.P_U$ represents the set of user-defined properties describing the entity e_i. The set $D_P = \{e_i.P_P\colon i = 1, ...n\}$, where $e_i.P_P$ represents the set of primitive properties describing the entity e_i, can be used to validate the results of the algorithm and to specify type checking rules. This validation process is out of the scope of this paper.

4 Using DBscan for Clustering Linked Data

To identify groups of similar entities, a density-based clustering algorithm considers clusters as dense regions. An object is dense if the number of its neighbors exceeds a threshold. The algorithm tries to identify the clusters based on the density of objects in a region. The objects are not grouped on the basis of a distance but when the neighbor density exceeds a certain threshold. A density-based algorithm is suitable for linked data because it finds classes of arbitrary shape, it is robust to noise and deterministic. In addition, unlike the algorithms based on k-means [20] and k-medoid [21] the number of classes is not required.

DBscan is a density-based clustering algorithm. In our context, we are especially interested in the concept of accessibility in this algorithm. The concept of accessibility of DBscan allows to group entities from one point to another considering both similarity and density. Consider ε the minimum similarity value for two entities to be considered as neighbors, V_ε (q) the set of neighbors of an entity q and $MinPts$ the minimum number of entities in a neighborhood. DBscan is based on the following definitions [15]:

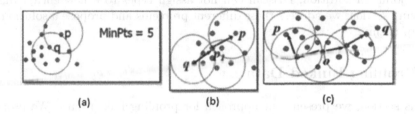

Fig. 2. Illustration of "reachability" (a), "accessibility" (b) and "cluster" (c) [15].

Definition 3 (Reachability). An object p is directly density-reachable from q (Fig. 2(a)) if:

- $p \in V_\varepsilon(q)$;
- $|V_\varepsilon(q)| \geq MinPts$ (q is a core).

Definition 4 (Accessibility). An object p is accessible from q (Fig. 2(b)) if there are n objects $p_1, ..., p_n$, with $p_1 = q$, $p_n = p$, such that p_{i+1} is directly density-reachable from p_i for $1 < i < n$.

Definition 5 (Cluster). A cluster is the maximal set of connected objects considering that an object p is connected to q if there is an object o such that p and q are both accessible from o (see Fig. 2(c)).

The concept of accessibility in DBscan allows (i) to cluster data without having to set the number of clusters because it is a clustering propagation; (ii) to detect noise, because the objects which are not accessible are considered as noisy; (iii) to cluster data in a single iteration in a deterministic way and discover clusters with arbitrary shape, which is important for linked data where similar entities may have heterogeneous property set.

To maximize the accessibility we need to maximize the reachability by minimizing the parameter $MinPts$. It represents the minimum number of entities in the neighborhood required for an entity to be a core and, in our context, to generate a type; it allows to exclude the outliers and the noise. As two entities of the same type can be described by different property sets, we consider in our approach that it is sufficient that an entity is similar to another entity to be considered as not outlying and to form a type. For this reason and to maximize the accessibility, we set $MinPts$ to 1. This implies that each object in the dataset is a core if it has at least one neighbor, and it is only required that two objects p and q are similar to be reachable from each other. We can say that, in our setting and with these parameters values, the DBscan algorithm is more accessibility-based than density-based.

DBscan requires the similarity threshold parameter for the clustering which is very important for the quality of the resulting types. This parameter is not obvious to set. Another requirement that DBscan does not meet in our context is that it can not assign several types to an entity because the discovered clusters are disjoint. In addition, DBscan can not assign types to a new entity. In the following section, we address these different problems and propose a solution for each of them.

5   Profiling Linked Data

In this section, we present our approach for profiling linked data. We propose in Subsect. 5.1 a self-adaptive profiling algorithm and show how we automatically detect the similarity threshold. In Subsect. 5.2, we propose an algorithm detecting overlapping types and allowing multiple typing of entities. As the dataset evolves and new entities are added, Subsect. 5.3 is devoted to incremental profiling.

5.1   Self-Adaptive Profiling

DBscan does not require the number of clusters. However, it requires the value of the similarity threshold ε, representing the minimum similarity value for two entities to be considered as neighbors. This parameter is not easy to set: if we choose a low similarity threshold, entities of different types may be grouped together and noisy entities may be incorrectly assigned to a cluster (see Fig. 3(b)); in contrast, if we choose a high similarity threshold, entities of the same type are not necessarily grouped together and entities which are not noisy are not assigned to a cluster (see Fig. 3(c)). As shown in Fig. 3(a), an optimal similarity threshold is the similarity value that permits to group entities of the same nature and to exclude noise.

Figure 4 gives the intuition behind our approach to automatically detect the similarity threshold. We can observe that for the entities that are grouped together, the distance between an entity and its nearest neighbor is small. However, the distance between an entity which is an outlier and its nearest neighbor is important. We can consider that the smallest distance between an outlier and its nearest neighbor gives an idea about the similarity threshold. In other words,

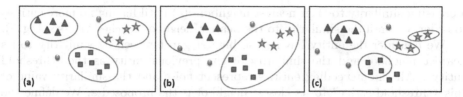

Fig. 3. Clustering results with different similarity thresholds.

the highest similarity between an entity which is not outlier and its nearest neighbor is the similarity threshold. The problem is how to identify this border? In our approach, we try to determine when the distance between an entity and its nearest neighbor becomes important to detect the similarity threshold for a dataset.

To measure the similarity between two property sets $e.P_U$ and $e'.P_U$ describing two entities e and e' respectively, we use Jaccard similarity. It is the ratio between the size of the intersection of the considered property sets and the cardinality of their union which is adapted to asses the similarity between the sets of different size. To detect when the distance between an entity e and its nearest neighbor becomes important, we propose to order the entities according to their

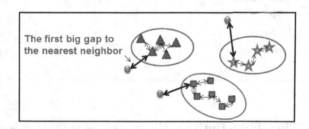

Fig. 4. Finding the similarity threshold from the nearest neighbor.

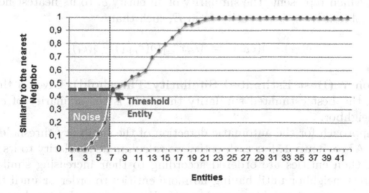

Fig. 5. Automatic detection of the similarity threshold.

increasing similarity to their nearest neighbor $\delta(e)$ until having no more entities to order, or until the similarity to the nearest neighbor reaches "1" (see Fig. 5).

We consider the entity having the biggest gap between its similarity to its nearest neighbor and the similarity of the previous entity as the "threshold entity". All the preceding entities represent noise and the similarity value of this "threshold entity" to its closest neighbor is the proposed ε. We define the "threshold entity" and the "best estimated similarity threshold" as follows.

Algorithm 1. Automatic Detection of the Similarity Threshold

Require: D_U
  $maxGap = 0$
  for all $e_i.P_U \in D_U$ **do**
    Calculate the similarity to its nearest neighbor as $\delta(e_i.P_U)$
  end for
  while ($\exists$ "not marked" $e_i.P_U \in D_U \wedge \delta(e_i.P_U)! = 1$) **do**
    Order $e_i.P_U$ in $orderdEntities$ according to increasing similarity to their first nearest neighbor
    mark $e_i.P_U$
  end while
  while $\exists$ $e_{j+1}.P_U \in orderdEntities$ **do**
    $gap = \delta(e_{j+1}.P_U) - \delta(e_j.P_U)$
    if $maxGap < gap$ **then**
      $maxGap = gap$
      $\varepsilon = \delta(e_{j+1}.P_U)$
    end if
  end while
  return ε

Definition 6 (Threshold Entity). Given a set of entities ordered according to their increasing similarity to the nearest neighbor $E = \{e_1, ..., e_i, e_{i+1}, ...e_n\}$, and $\delta(e_i)$ which represents the similarity of an entity e_i to its nearest neighbor, the threshold entity e_t is an entity from E, such that:

$$\delta(e_t) - \delta(e_{t-1}) = Max_{i=1}^{n}(\delta(e_{i+1}) - \delta(e_i)) \tag{1}$$

Definition 7 (Best Estimated Similarity Threshold). Given a threshold entity e_t, the best estimated similarity threshold is the similarity of e_t to its nearest neighbor.

Our approach for the automatic detection of the similarity threshold is presented in **Algorithm 1**. For each entity we calculate its similarity to its nearest neighbor, then entities are ordered according to their increasing similarity to their nearest neighbor until having no more entities to order or until the similarity to the nearest neighbor reaches "1". We compute the gap between the similarity to the nearest neighbor of each successive entities. We consider the similarity to the nearest neighbor of the entity having the biggest gap with the similarity of the previous entity as the similarity threshold ε.

The complexity of calculating the similarity between each entity pair to find the best neighbor and thus detecting the optimal similarity threshold is $O(n^2)$, where n is the number of entity. However, the calculation of the similarity of distinct entities to all the others can be parallelized. This reduces the complexity for each entity to $O(n)$, with n parallel processing. The complexity of DBscan is $O(n^2)$, for a dataset with n entities. In [15], an extension to the algorithm is proposed to reduce the complexity of DBscan to $O(n * log(n))$ by using spatial access methods such as R*-trees [5].

5.2   Detecting Overlapping Profiles

DBscan provides a set of disjoint clusters, but an entity may have several types. Our problem is how to assign several types to an entity? Our intuition is that an entity has a given type if it is described by all the strong properties of this type. By strong property, we mean that the probability of the property in the entities of the type is high. We have adapted DBscan to build a profile during the clustering process for each discovered type. The result is a set of disjoint clusters with their associated profiles. We generate overlapping clusters by analyzing these type profiles. We define a type profile as follows.

Definition 8 (Type Profile). A type profile is a property vector where each property is associated to a probability. The profile corresponding to a type T_i is denoted $TP_i = ((p_{i1}, \alpha_{i1}), ..., (p_{in}, \alpha_{in}))$, where each p_{ij} represents a property and where each α_{ij} represents the probability for an entity of T_i to have the property p_{ij}.

The type profile represents the canonical structure of the type T_i. The probability α_{ij} associated to a property p_{ij} in the profile of type T_i is evaluated as the number of entities in T_i having the property p_{ij} over the total number of entities in T_i. We define a strong property as follows.

Definition 9 (Strong Property). A property p is a strong property for a type T_i given a threshold θ, if (p, α) $\in TP_i$ and $\alpha \geq \theta$.

Figure 6 shows an example of clusters and the corresponding type profiles. Considering that $\theta = 1$, we can see that TP_1 has all the strong properties of the "Author" profile, which means that entities of TP_1 also have the type "Author". We can also observe that TP_1 has all the strong properties of the "Organizer" profile, which means that its entities also have the type "Organizer". In the same way, we compare all the pairs of profiles provided by our algorithm, and each time all the strong properties of a profile TP_i are found in another profile TP_j, then the type T_i is added to the entities of the type T_j.

The importance of a property for a given type is captured by the associated probability. We generate overlapping clusters using type profiles as described in **Algorithm 2**. We compare each pair of type profiles to detect the possible inclusion of strong properties. Considering the type T_i described by the profile

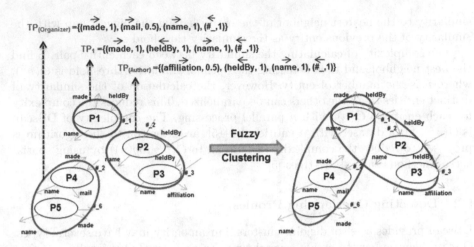

Fig. 6. Overlapping clustering.

Algorithm 2. Overlapping Profiling

Require: *typeProfileSet*
  for all $TP_i \in typeProfileSet$ **do**
    for all $TP_j \in typeProfileSet$ with $i! = j$ **do**
      if $\forall (p, \alpha)$ in TP_i, $\alpha \geq \theta$: (p, α) in TP_j **then**
        Add T_i to types of the entities of the cluster j
      end if
    end for
  end for

TP_i, if every strong property p of TP_i belongs to another type profile TP_j, then the type T_i falls within the types of entities of the cluster j.

Algorithm 2 compares each pair of type profiles, its complexity is $O(k^2)$, where k is the number of types in a cluster. Given n the number of entities in the dataset, $k < n/2$, because the minimum number of entities to form a type is 2.

5.3 Incremental Profiling

Linked datasets can evolve and new entities may be added over a period of time. Given a set of clusters, the problem we are interested in is how to assign a new entity to these existing clusters? A new entity can belong to one or several clusters; it can also form a new cluster with an unclassified entity, or it can be considered as outlying. An entity is considered as outlying and classified as noise if it has no neighbors. If a new entity is similar to other outlying entities in the dataset, they will form a new cluster with all similar entities. In our context, the clusters reflect the types of the dataset. A dataset focuses generally on a specific area such as the conference dataset which contains data about articles, authors, organizers, etc. We consider that the discovered types in the dataset do not have to be redefined each time the dataset evolves. However, new types can be generated. Therefore, incremental profiling should be more oriented on typing new entities than on redefining of the existing clusters. Our goal is to assign the new entity to one or several existing clusters and possibly generate a new cluster from this entity.

Ester et al. have proposed an incremental version of DBscan in [14]. They have identified the set of impacted objects by the inserted/deleted entity and reapplied DBscan on this set. The intermediate results of the previous clustering are kept in memory, such as the connections between the objects. The main purpose is to monitor the evolution of the clusters. However, it can not deal with memory limitations as it stores all the intermediate results of the previous clustering. In addition, similarly to DBscan, it can not assign several types to an entity and the similarity threshold is required. Deleting an entity may lead to the disappearance of a cluster of entities which will then become noise. In our context, a cluster reflects a type. After the discovery of a type in a dataset, it is not desirable to lose the information of the existence of this type in the dataset. Similarly, entities that were previously typed should not be considered as noise when an entity is deleted, because similar entities could be inserted later. For these reasons, we do not update the clusters when an entity is deleted. The insertion of a new entity may cause the merge of two clusters, which in our case corresponds to the loss of two types, which is not desirable. Instead of merging the two clusters, we will assign the entity to both of them. Beside the insertion or the deletion of entities, another evolution which may occur is the modification of the property set of an entity; we consider that this is equivalent to a deletion followed by an insertion.

To find the type(s) of a new entity e we can use a data mining classification algorithm, such as K-NN [12]. However, this algorithm requires the number of nearest neighbors K. We can not specify this parameter, because we can not predict the number of types for an entity. In addition, this algorithm is costly as we need to search the nearest neighbors in the whole dataset.

To classify a new entity e, we propose to generate a fictive nearest neighbor b called "best fictive neighbor" of e from each type profile TP in the dataset. The best fictive neighbor of e from TP is composed of the common properties between e and TP and the strong properties of TP. We define the best fictive neighbor as follows.

Definition 10 (Best Fictive Neighbor). The best fictive neighbor b of an entity e considering a type profile TP is a fictive entity described by $b.P_U$ a set of user defined properties, such as:

- If $p \in e.P_U$ and $(p, \alpha) \in TP$, then $p \in b.P_U$;
- If $(p, 1) \in TP$, then $p \in b.P_U$.

Algorithm 3. Incremental Multiple Classification for an Entity

---

Require: new entity e, type profiles $\{TP_{class}\}$, set of outlying entity N, similarity
 threshold ϵ
  $A = e$
  for each TP_{class} **do**
     Extract the best fictive neighbor b for e from TP_{class}
     if Similarity(b,e) $>= \epsilon$ **then**
        Add the type of TP_{class} to e
     end if
  end for
  for each $o \in N$ **do**
     if Similarity(o,e) $>= \epsilon$ **then**
        group o and e into A
     end if
  end for
  if e has types assigned **then**
     Add the types of e to each $o \in A$
     if e has only one type T_{class} **then**
        for each $r \in A$ **do**
           Update the profile TP_{class}
        end for
     end if
  else
     if $|A| > 1$ **then**
        A represents a new type
        Create a new profile for the type A
     else
        e is an outlying entity
     end if
  end if

We compute the similarity between the best fictive neighbor b and the new
entity e in the same way as for entities during the clustering process, by adapting
the Jaccard similarity. In order to measure the similarity between two property
sets $b.P_U$ and $e.P_U$ describing two entities b and e respectively, we use Jaccard
similarity defined as follows:

$$Similarity(b, e) = \frac{|b.P_U \cap e.P_U|}{|b.P_U \cup e.P_U|} \tag{2}$$

We consider the two entities b and e as neighbors if the similarity between them
is greater than the similarity threshold ϵ determined as in Sect. 5.1. In this case,
we assign the type of the profile from where the best fictive neighbor b was
extracted, to e. We repeat this process for each type profile in the dataset as
it is shown in Algorithm 3, which has a complexity of $O(k)$ where k represents
the number of types in the dataset. If there are outlying entities in the dataset
which are similar to the new entity, we assign the discovered types of the new
entity to these outlying entities. However, if there are no types discovered for

the new entity, we build a new type with similar outlying entities. If e has no similar best fictive neighbor and no similar outlying entities, it is considered as an outlier. We update a type profile only if the entity e has one type, because if we consider entities having several types in the type profiles we might introduce properties characterizing another type.

Each time the type of a new entity is discovered, the corresponding profile is updated using the **UpdateTypeProfile** function (**Algorithm 4**), which adds the properties not already in the profile and recomputes the probabilities of the existing ones. This function is also used to build the profiles of the types during the clustering process.

Algorithm 4. UpdateTypeProfile

Require: TP_{class}, $class$, $e.P_U$, $nbEntitiesInClass$
   for each property $p \in r.P_U$ **do**
      if p is in TP_{class} with α its probability **then**
        $\alpha \leftarrow \frac{(\alpha \times nbEntitiesInClass)+1}{nbEntitiesInClass+1}$
        update α in TP_{class}
      else
        $TP_{class} \leftarrow TP_{class} \cup (p, 1/(nbEntitiesInClass + 1))$
      end if
   end for

---

6 Evaluation

This section presents some experimentation results using our approach. We have generated the similarity threshold, then we have evaluated the proposed similarity threshold considering the number of generated clusters, the percentage of noise and the quality of the generated types.

For our experiments, we have used three datasets: the Conference[6] dataset, which exposes data for several Semantic Web conferences and workshops with 11 classes and 1430 triples; the BNF[7] dataset which contains data about the French National Library (Bibliothèque Nationale de France) with 5 classes and 381 triples; we have extracted a dataset from DBpedia[8] with 19696 triples considering the following types: Game, Stadium, Footballer, Politician, Film, Museum, Book and Country. The tests have been performed on an Intel(R) Xeon(R) machine, CPU of 2.80 GHz, 64-bit with 4 GB of RAM. We have used well-known information retrieval metrics described in the following section.

---

[6] Conference: data.semanticweb.org/dumps/conferences/dc-2010-complete.rdf.
[7] BNF: datahub.io/fr/dataset/data-bnf-fr.
[8] DBpedia: dbpedia.org.

6.1   Metrics and Experimental Methodology

In order to evaluate the similarity threshold proposed by our algorithms, we vary the threshold value and we compare: (i) the number of resulting clusters with the number of types in the dataset, (ii) the percentage of detected noise and (iii) the quality of the generated types. To evaluate the quality of the generated types, we have extracted the existing type definitions from our datasets and considered them as a gold standard. We have then run our algorithms on the dataset without the type definitions, with different similarity thresholds, and evaluated the precision and recall for the inferred types. We have annotated each inferred cluster C_i with the most frequent type label associated to its entities. For each type label L_i corresponding to type T_i in the dataset and each cluster C_i inferred by our algorithm, such that L_i is the label of C_i, we have evaluated the precision $P_i(T_i, C_i) = |T_i \cap C_i|/|C_i|$ and the recall $R_i(T_i, C_i) = |T_i \cap C_i|/|T_i|$. We denote k the number of generated clusters; to assess the overall type precision and recall, each type is weighted according to its number of entities as follows:

$$P = \sum_{i=1}^{k} \frac{|C_i|}{n} \times P_i(T_i, C_i) \qquad\qquad R = \sum_{i=1}^{k} \frac{|C_i|}{n} \times R_i(T_i, C_i)$$

To show that our approach can be used to classify the new incoming entities in the dataset, we have used a cross-validation training technique by dividing randomly each dataset into a training set ($2/3$ of the dataset) and a test set ($1/3$ of the dataset). We have applied our self-adaptive approach on the training set to build the type profiles. We have applied our incremental profiling approach on the entities of the test set to find the appropriate types for each entity. The precision of the approach is the percentage of correctly classified entities in the test set according to their declared types in the dataset. The recall of the approach is the percentage of entities of a given type that are not found in the test set. Given TP the number of true positives, FP the number of false positives and FN the number of false negatives, we evaluate the precision P, the recall R and the $F1$ score as follows:

$$R = \frac{TP}{TP + FN}$$

$$P = \frac{TP}{TP + FP} \qquad\qquad F1 = \frac{2TP}{2TP + FP + FN}$$

6.2   Results

We have automatically detected the similarity threshold ε for each dataset. Figure 7 represents entities of each dataset ordered according to the decreasing values of their similarity to their nearest neighbor. We have not represented all the entities of the dataset, because the similarity to the nearest neighbor

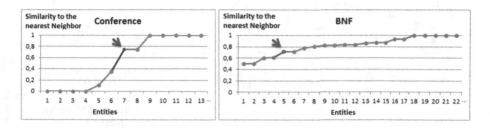

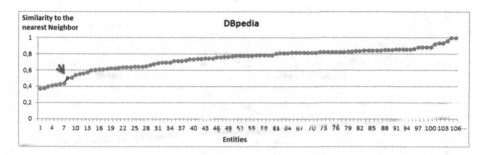

Fig. 7. Automatic detection of similarity threshold.

remains the same when it reaches 1. We consider the entity having the biggest gap with the similarity of the previous entity as the "threshold entity". The entities on the left of the "threshold entities" are considered as noisy. The values of the best similarity of each "threshold entity" represent the ε of each dataset according to the distribution of their entities as follows: $\varepsilon = 0.75$ for the Conference dataset; $\varepsilon = 0.72$ for the BNF dataset and $\varepsilon = 0.5$ for the DBpedia dataset. The entities of the same type, in the Conference and BNF datasets, have more homogeneous property sets than the entities in DBpedia, which explains why the detected similarity threshold is lower for the DBpedia dataset. Calculating the similarity between each pair of entities has a complexity of $O(n^2)$, and the running time for each dataset is 16 milliseconds (ms) for the Conference dataset, 2 ms for the BNF dataset and 32 ms for the DBpedia dataset. The automatic detection of the similarity threshold takes a few milliseconds for each dataset because it has a linear complexity $O(n)$ as the similarity between an entity and its nearest neighbor is calculated beforehand.

Figure 8 shows the number of clusters (a), the percentage of noise (b) and the precision/recall of the generated types according to the similarity threshold in the Conference dataset. The results for the BNF and DBpedia datasets are given in Figs. 9 and 10 respectively. The vertical dashed line represents the similarity threshold proposed by our approach. For the Conference dataset (see Fig. 8), the number of clusters increases when increasing the similarity threshold. Indeed, the higher the similarity threshold, the more entities tend to be grouped in

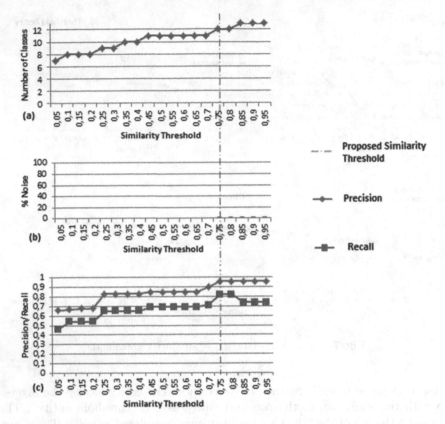

Fig. 8. Number of clusters (a), Percentage of noise (b) and the Precision/Recall of the generated types according to the similarity threshold in conference dataset.

different clusters. The similarity threshold determined by our approach gives a correct number of clusters which is equal to the number of types of the dataset. The Conference dataset contains a very low percentage of noise (2.4 %) and the precision and the recall are both quite high, especially with the generated similarity threshold. The percentage of noise stays stable even with a very high value of the similarity threshold (0.95), because the entities of the same type in the Conference dataset have very homogeneous property set.

For the BNF dataset (see Fig. 9), the similarity threshold proposed by our approach gives a correct number of clusters which is equal to the number of types of the dataset. The precision and recall vary depending on the similarity threshold. The precision increases when the similarity threshold increases. A precision of 1 is achieved with the similarity threshold proposed by our approach. However the recall decreases then increases when increasing the similarity threshold, because if the threshold is high, instances of a given type are considered noisy while they are not. Indeed, the percentage of detected noise increases when increasing the similarity threshold until reaching a percentage of 50 %,

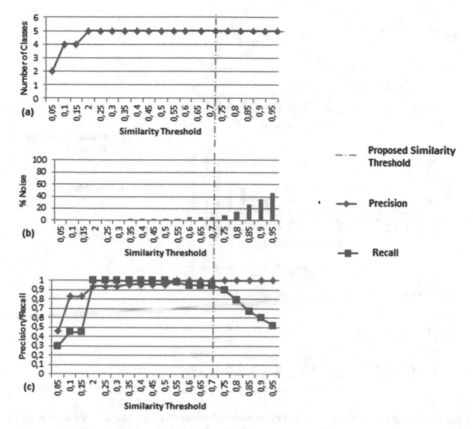

Fig. 9. Number of clusters (a), Percentage of noise (b) and the Precision/Recall of the generated types according to the similarity threshold in BNF dataset.

and in this case, the recall decreases because typed instances in the dataset were not grouped in clusters as the similarity threshold is too high. The similarity threshold proposed by our approach gives a good recall with a reasonable noise percentage. The recall is not maximum because some entities are considered as noisy because they are very different from the other entities. It is the case for some entities of the type "Expression" which are considered as noisy but typed as "Expression" in the dataset.

For the DBpedia dataset (see Fig. 10), the number of clusters varies according to the similarity threshold. In principle, the higher the similarity threshold, the higher the number of clusters. However, for this dataset, instances that are not noisy are considered noisy when the similarity threshold is too high, which has reduced the number of clusters. The similarity threshold proposed by our approach gives a correct number of clusters which is equal to the number of types of the dataset. The precision increases as the similarity threshold increases to achieve a precision of 1, which is also the precision obtained with the similarity threshold proposed by our approach. However, the recall decreases then

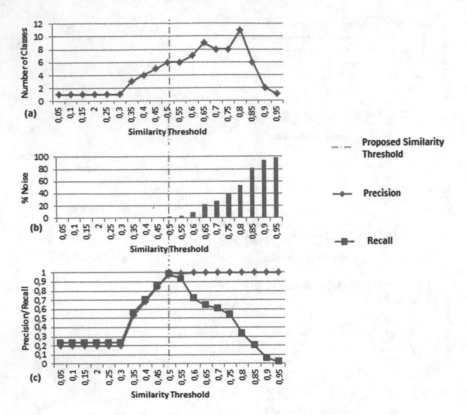

Fig. 10. Number of clusters (a), Percentage of noise (b) and the Precision/Recall of the generated types according to the similarity threshold in the DBpedia dataset.

increases when increasing the similarity threshold, because when the threshold is high, instances of a given type could be considered noisy when they are not. Indeed, the percentage of detected noise increases when the similarity threshold increases until reaching a percentage of 98 %, and in this case, the recall decreases because some typed instances in the dataset were not grouped in clusters, as the similarity threshold was too high. The threshold proposed by our approach gives a good recall with reasonable noise percentage.

The entities of the same type in the Conference dataset have homogeneous property sets, therefore, only a little noise is detected even when the similarity threshold is very high (0.95). However, there are properties shared between the property sets of entities of different types, such as the "name" property between the types "Organizer" and "Author", and the "based-near" property between the types "City" and "Point". Therefore, the number of classes varies according to the similarity threshold more than for the BNF dataset, where the number of classes remains the same and is equal to 5 as soon as the similarity threshold reaches 2. Indeed, the entities of different types in the BNF dataset share very few properties. However, the property sets of entities of the same type are less homogeneous than for the Conference dataset. Therefore, the noise is more important when the similarity threshold increases. The entities of the DBpedia dataset are described by a large number of properties (average 150 property by type). Some of these properties are not specific to a type, such as "hasPhoto-Collection", which implies that entities of different types share many properties. Therefore, the number of classes is 1 when the similarity threshold is lower than 0.3. In addition, the property sets of entities of the same type are very heterogeneous unlike for the Conference and BNF datasets. Therefore, the detected similarity threshold is lower and the noise is much higher when the similarity threshold increases than for the Conference and BNF datasets. Given n the number of entity in a dataset, clustering each dataset takes a few seconds. Its complexity is $O(n * log(n))$ and the similarity between an entity and its nearest neighbor is calculated beforehand.

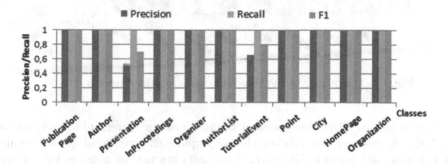

Fig. 11. The quality of the classified entities of conference dataset.

Figure 11 shows the precision and the recall of our incremental approach on the Conference test set. We can see that our approach achieves a good quality for the classified entities. The types "Presentation" and "TutorialEvent" do not have a good precision because they contain entities having different types in the dataset. However, these types have the same structure and it is therefore impossible to distinguish between them. Our algorithm has made the distinction between the types "City" and "Point" although they are very similar, as it did for the types "Author" and "Organizer". For entities having both the types "Author" and "Organizer", the approach typed them correctly.

Figure 12 shows the precision and the recall of our incremental approach on the BNF test set. We can see that our approach gives a good quality for the classified entities. The type "Expression" does not have a good precision because some of its entities are considered as noisy. These entities are typed as "Expression" in the dataset, however, they are very different than the other entities of this type, therefore, they are considered as outlying entities. The types of the BNF dataset are quite dissimilar, which means that their profiles do not share many common properties. This explains the quality of the classification.

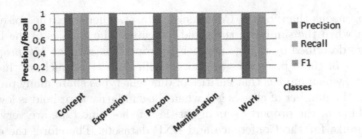

Fig. 12. The quality of the classified entities of BNF dataset.

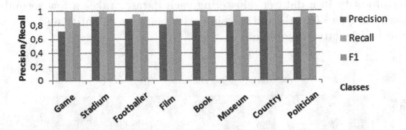

Fig. 13. The quality of the classified entities of the DBpedia dataset.

Figure 13 shows the precision and the recall of our incremental approach on the DBpedia test set. We can see that our approach gives good quality results for the classified entities. However, the results are not as good as for Conference and BNF datasets, because there are many shared properties between the type profiles. In addition, the average number of properties for an entity is 150. These entities have general properties that are not specific to the type, such as wikiPageID, hasPhotoCollection, wikiPageDisambiguates, primaryTopic, has-PhotoCollection, etc. There are some entities for a given type that do not have the specific properties of this type, which implies that they are detected as noisy instances by our approach, such as for instances of the "Footballer" type, which do not have any specific property of this type, such as team or clubs.

The classification process for each dataset takes a few milliseconds. It has a linear complexity of $O(k)$, where k is the number of types in each dataset. The number of types in each dataset is generally very low compared to the number of entities. The number of types is respectively 11, 5 and 8 for the Conference, the BNF and the DBpedia datasets.

7   Related Works

To facilitate the exploration and reuse of existing RDF datasets, descriptive and reliable metadata is required. However, as witnessed in the popular dataset registry DataHub[9], dataset descriptions are often missing entirely, or are

---

[9] http://datahub.io/group/lodcloud.

outdated [9]. Our work can complement the existing approaches for profiling a dataset described in [7,17]. Indeed, the information in our type profiles can be used to profile the whole dataset by generating VoID[10] data. Profiling a dataset consists generally in extracting metadata information about the dataset, such as the number of types in the dataset, the number of instances for each type, the number of instances having a given property, etc.

Some works address the problem of discovering the types of a dataset, which consists mainly in discovering clusters of similar entities to form a type. In [11,36], an approximate DataGuide based on COBWEB [18] is proposed. The resulting clusters are disjoint, and the approach is not deterministic. Nestorov et al. [29] use bottom-up grouping providing a set of disjoint clusters, where the number of clusters is required. Christodoulou et al. [10] propose the use of standard ascending hierarchical clustering to build structural summaries of linked data. Each instance is represented by its outgoing properties and the property set of a cluster is the union of the properties of its entities, while in our approach, the probability of each property is computed for a type. The algorithm provides disjoint clusters; the hierarchical clustering tree is explored to assess the best cutoff level, which could be costly. ProLOD [2,8] proposes to profile a dataset by applying the k-means algorithm with different k values, representing different number of clusters, which could also be costly. Fanizzi et al. [16] propose to group instances into clusters on the basis of their different types. In other words, instances that share many common types will be assigned to the same cluster. A genetic algorithm [27] is used to find the best centers. Unlike our approach, the purpose of this method is to group instances sharing the same types and not to find new types of instances. Indeed, type information is required because the clustering is based on the types declared for a given instance, whereas in our approach, it is based on the property set of an instance. The approach does not require the number of clusters as a parameter. However, it starts with a large number of representatives of the population and reduces them to find the optimal number, which could be expensive.

The authors of DBscan [15] have proposed to discover the similarity threshold by detecting the noisy points depending on the distribution of the dataset density. The number of neighbors $MinPts$ is set to 4. The function $4 - dist(p)$ represents the distance between a point p in the dataset and its fourth neighbor. The points of the dataset are ordered according to the decreasing values of their similarity. The first valley of the graph represents the value of the similarity threshold. If the percentage of noise in the dataset is known, we can consider this percentage to find the point of the similarity threshold. However, the first valley is hard to find because the percentage of noise in the dataset is generally unknown, and the automatic detection of the first valley of the graph is not obvious. Another approach, OPTICS [3], has proposed a solution to find the best similarity threshold for the DBscan algorithm. It consists in performing DBscan for different values of similarity threshold, then choose the smallest threshold that provides the highest density.

---

[10] http://vocab.deri.ie/void.

The problem of choosing an adequate value for the similarity threshold has been the focus of several research works but it remains an open issue. An automatic method to determine a similarity threshold for generating a well-defined clustering for large datasets is proposed in [34], where the similarity threshold is increased with a small constant, and several clusterings are generated. The method calculates for each clustering, the intra-cluster similarity and provides the threshold that generates the minimum value. The method is very costly because the clustering is re-run for each similarity value. The increase of the similarity threshold is set manually or calculated according to the number of attributes. However, it does not ensure to have a different clustering for each iteration, and in our context, the number of attribute varies from one instance to another. In the approach described in [30], a hierarchy of data is generated using hierarchical clustering, the user is asked to choose the similarity threshold according to the number of clusters. The method does not determine automatically the value of the similarity threshold, and this technique was not developed in order to process large datasets. A similar work is presented in [32], where the similarity threshold is replaced by the number of clusters to be generated. However, in our case the number of types in the dataset is unknown.

Data mining classification algorithms can assign a type to a new entity, such as K-NN [12] as it is used in [33]. However, this algorithm needs some parameters such as the number of nearest neighbors K. We can not specify this parameter, because we do not know the number of types for an entity. In addition, this algorithm is costly as we need to search the nearest neighbors in the whole dataset. An incremental version of DBscan is presented in [14]. The general strategy for updating a clustering is to start the DBscan algorithm only with core objects that are in the neighborhood of the deleted or inserted object. It is not necessary to rediscover density-connections which are known from the previous clustering and which are not changed by the update operation. However, the approach can not deal with memory limitations as it stores all the intermediate results of the previous clustering. In addition, similarly to DBscan, it can not assign several types to an entity and the similarity threshold is required.

8   Conclusion

In this paper, we propose a self-adaptive and incremental approach for profiling linked data. Our work focuses on the data mining task of clustering to profile the dataset and to discover the overlapping types. Our approach does not require the number of types in the dataset, it considers the evolution of a dataset and it can assign several types to an entity. We propose a self-adaptive method to automatically detect the similarity threshold according to the distribution of the similarity of the entities to their nearest neighbor. We propose to build a profile for each cluster during the process of clustering to summarize its content. We use these profiles to discover overlapping clusters which allows to assign several types to an entity. To type a new entity without browsing the whole dataset, we use the profiles by generating the best fictive neighbor for the new entity.

Our experiments show that our approach detects a good similarity threshold on different datasets, considering the real number of types, the percentage of noise and the quality of the results according to the precision and the recall. In addition, the approach gives good quality results for incremental profiling. Our approach helps to understand the content of a dataset by providing a high-level summary defined according to the types contained in this dataset. This is useful for querying a dataset, as it provides the properties and types existing in this dataset. The profiles could also be used for interlinking and matching datasets. Indeed, matching the types first, when they are known, could reduce considerably the complexity of the process.

In our future works, we will address the annotation of the inferred types. Indeed, in addition to identifying a cluster of entities having the same type, it is also useful to find the labels which best capture the semantics of this cluster. Another important problem we are interested in is how to use the profiles to decompose a query and sending the sub-queries to the relevant sources, when accessing multiple distributed datasets. The probabilities of the properties could be used to optimize the execution plans by ordering the sub-queries according to the selectivity of their criteria.

Acknowledgments. This work was partially funded by the French National Research Agency through the CAIR ANR-14-CE23-0006 project.

References

1. The World Wide Web Consortium (w3c) - RDF 1.1 concepts and abstract syntax. https://www.w3.org/TR/rdf11-concepts/
2. Abedjan, Z., Gruetze, T., Jentzsch, A., Naumann, F.: Profiling and mining RDF data with ProLOD++. In: 2014 IEEE 30th International Conference on Data Engineering (ICDE), pp. 1198–1201. IEEE (2014)
3. Ankerst, M., Breunig, M.M., Kriegel, H.-P., Sander, J.: Optics: ordering points to identify the clustering structure. ACM Sigmod Record **28**, 49–60 (1999). ACM
4. Auer, S., Bizer, C., Kobilarov, G., Lehmann, J., Cyganiak, R., Ives, Z.: DBpedia: A nucleus for a web of open data the semantic web. The Semantic Web (2007)
5. Beckmann, N., Kriegel, H.-P., Schneider, R., Seeger, B.: The R*-tree: an efficient and robust access method for points and rectangles. ACM SIGMOD Record **19**, 322–331 (1990). ACM
6. Bezdek, J.C., Ehrlich, R., Full, W.: FCM: The fuzzy C-Means clustering algorithm. Comput. Geosci. **10**, 191–203 (1984)
7. Böhm, C., Lorey, J., Naumann, F.: Creating void descriptions for web-scale data. Web Semant. Sci. Serv. Agents World Wide Web **9**(3), 339–345 (2011)
8. Böhm, C., Naumann, F., Abedjan, Z., Fenz, D., Grütze, T., Hefenbrock, D., Pohl, M., Sonnabend, D.: Profiling linked open data with ProLOD. In: 2010 IEEE 26th International Conference on Data Engineering Workshops (ICDEW), pp. 175–178. IEEE (2010)
9. Buil-Aranda, C., Hogan, A., Umbrich, J., Vandenbussche, P.-Y.: SPARQL web-querying infrastructure: ready for action? In: Alani, H., et al. (eds.) ISWC 2013. LNCS, vol. 8219, pp. 277–293. Springer, Heidelberg (2013). doi:10.1007/978-3-642-41338-4_18

10. Christodoulou, K., Paton, N.W., Fernandes, A.A.A.: Structure inference for linked data sources using clustering. In: Hameurlain, A., Küng, J., Wagner, R., Bianchini, D., Antonellis, V., Virgilio, R. (eds.) Transactions on Large-Scale Data- and Knowledge-Centered Systems XIX. LNCS, vol. 8990, pp. 1–25. Springer, Heidelberg (2015). doi:10.1007/978-3-662-46562-2_1
11. Clerkin, P., Cunningham, P., Hayes, C.: Ontology discovery for the semantic web using hierarchical clustering. Semantic Web Min. **1**, 17 (2001)
12. Dasarathy, B.V.: Nearest neighbor (NN) norms: NN pattern classification techniques (1991)
13. Dempster, A.P., Laird, N.M., Rubin, D.B.: Maximum likelihood from incomplete data via the EM algorithm. J. Royal Stat. Soc. **39**, 1–38 (1977)
14. Ester, M., Kriegel, H.-P., Sander, J., Wimmer, M., Xu, X.: Incremental clustering for mining in a data warehousing environment. In: Proceedings of the 24th International Conference on Very Large Data Bases, pp. 323–333. Morgan Kaufmann Publishers Inc. (1998)
15. Ester, M., Kriegel, H.-P., Sander, J., Xu, X.: A density-based algorithm for discovering clusters in large spatial databases with noise. In: KDD (1996)
16. Fanizzi, N., Amato, C.D., Esposito, F.: Metric-based stochastic conceptual clustering for ontologies. Inf. Syst. **34**(8), 792–806 (2009)
17. Fetahu, B., Dietze, S., Pereira Nunes, B., Antonio Casanova, M., Taibi, D., Nejdl, W.: A scalable approach for efficiently generating structured dataset topic profiles. In: Presutti, V., d'Amato, C., Gandon, F., d'Aquin, M., Staab, S., Tordai, A. (eds.) ESWC 2014. LNCS, vol. 8465, pp. 519–534. Springer, Heidelberg (2014). doi:10.1007/978-3-319-07443-6_35
18. Fisher, D.H.: Knowledge acquisition via incremental conceptual clustering. Mach. Learn. **2**(2), 139–172 (1987)
19. Heath, T., Bizer, C.: Linked data: Evolving the web into a global data space. Synth. Lect. Semant. Web Theor. Technol. **1**(1), 1–136 (2011)
20. Jain, A.K.: Data clustering: 50 years beyond k-means. Pattern Recogn. Lett. **31**, 651–666 (2010)
21. Kaufman, L., Rousseeuw, P.: Clustering by Means of Medoids. Reports of the Faculty of Mathematics and Informatics, Faculty of Mathematics and Informatics (1987)
22. Kellou-Menouer, K., Kedad, Z.: Evaluating the gap between an RDF dataset and its schema. In: Jeusfeld, M.A., Karlapalem, K. (eds.) ER 2015. LNCS, vol. 9382, pp. 283–292. Springer, Heidelberg (2015). doi:10.1007/978-3-319-25747-1_28
23. Kellou-Menouer, K., Kedad, Z.: Schema discovery in RDF data sources. In: Johannesson, P., Lee, M.L., Liddle, S.W., Opdahl, A.L., López, Ó.P. (eds.) ER 2015. LNCS, vol. 9381, pp. 481–495. Springer, Heidelberg (2015). doi:10.1007/978-3-319-25264-3_36
24. Konrath, M., Gottron, T., Staab, S., Scherp, A.: Schemex: efficient construction of a data catalogue by stream-based indexing of linked data. Web Semant. Sci. Serv. Agents on the World Wide Web **16**, 52–58 (2012)
25. Li, H.: Data profiling for semantic web data. In: Wang, F.L., Lei, J., Gong, Z., Luo, X. (eds.) WISM 2012. LNCS, vol. 7529, pp. 472–479. Springer, Heidelberg (2012). doi:10.1007/978-3-642-33469-6_59
26. Maali, F., Campinas, S., Decker, S.: Gagg: a graph aggregation operator. In: Gandon, F., Sabou, M., Sack, H., d'Amato, C., Cudré-Mauroux, P., Zimmermann, A. (eds.) ESWC 2015. LNCS, vol. 9088, pp. 491–504. Springer, Heidelberg (2015). doi:10.1007/978-3-319-18818-8_30

27. Michalewicz, Z.: Genetic algorithms + data structures = evolution programs. Springer Science & Business Media (2013)
28. Naumann, F.: Data profiling revisited. ACM SIGMOD Record **42**(4), 40–49 (2014)
29. Nestorov, S., Abiteboul, S., Motwani, R.: Extracting schema from semistructured data. ACM SIGMOD Record **27**, 295–306 (1998). ACM
30. Pena, P.: Determining the similarity threshold for clustering algorithms in the logical combinatorial pattern recognition through a dendograme. In: 4th Iberoamerican Simposium of Pattern Recognition, pp. 259–265 (1999)
31. Quilitz, B., Leser, U.: Querying distributed RDF data sources with SPARQL. In: Bechhofer, S., Hauswirth, M., Hoffmann, J., Koubarakis, M. (eds.) ESWC 2008. LNCS, vol. 5021, pp. 524–538. Springer, Heidelberg (2008). doi:10.1007/978-3-540-68234-9_39
32. Reyes-Gonzalez, R., Ruiz-Shulcloper, J.: An algorithm for restricted structuralization of spaces. Proc. IV SIARP 267 (1999)
33. Rizzo, G., Fanizzi, N., d'Amato, C., Esposito, F.: Prediction of class and property assertions on OWL ontologies through evidence combination. In: Proceedings of the International Conference on Web Intelligence, Mining and Semantics, p. 45. ACM (2011)
34. Sánchez-Díaz, G., Martínez-Trinidad, J.F.: Determination of similarity threshold in clustering problems for large data sets. In: Sanfeliu, A., Ruiz-Shulcloper, J. (eds.) CIARP 2003. LNCS, vol. 2905, pp. 611–618. Springer, Heidelberg (2003). doi:10.1007/978-3-540-24586-5_75
35. Suchanek, F.M., Kasneci, G., Weikum, G.: Yago: a core of semantic knowledge. In: Proceedings of the 16th International Conference on World Wide Web (2007)
36. Wang, Q.Y., Yu, J.X., Wong, K.-F.: Approximate graph schema extraction for semi-structured data. In: Zaniolo, C., Lockemann, P.C., Scholl, M.H., Grust, T. (eds.) EDBT 2000. LNCS, vol. 1777, pp. 302–316. Springer, Heidelberg (2000). doi:10.1007/3-540-46439-5_21

Author Index

Printed in the United States
By Bookmasters